THE
NATURAL
WORLD
Chaos and
Conservation

THE
NATURAL
WORLD

Chaos and Conservation

CECIL E. JOHNSON
Associate Professor of Biology
Riverside City College

McGraw-Hill Book Company
New York St. Louis San Francisco Düsseldorf
Johannesburg Kuala Lumpur London Mexico
Montreal New Delhi Panama Rio de Janeiro
Singapore Sydney Toronto

This book was set in Palatino Alphatype by University Graphics, Inc., and printed and bound by The Murray Printing Company. The designer was J. E. O'Connor. The editors were Jeremy Robinson and Andrea Stryker-Rodda. Sally Ellyson supervised production.

The Natural World: Chaos and Conservation

Library of Congress Catalog Card Number 74-159308

1234567890MUMU7987654321

CONTENTS

Part 2

Introduction, 133

PREFACE

In 1962, the almost unnoticed wave which swelled to gigantic proportions on Earth Day, April 22, 1970, was begun with the publication of a book called "Silent Spring," written by a quiet, exacting scientist named Rachel Carson. Miss Carson's anguish overcame her personal distaste for publicity, and for the first time, a meaningful message forced scientist and nonscientist alike to take a long, hard look at chemicals such as DDT and their effects on our waters and the plants and animals of our environment.

Since then, other spokesmen have emerged from the ranks of silent scientists and quiet citizens to speak out against man's rape of the world, and for the future of mankind on the planet. The battle lines have been drawn. The war against pollution has begun. "The dirty animal — man" has become aware of, become concerned about, and at length begun to act on his suicidal tendencies toward the destruction of his planet.

Although the vested interests of commercialism and some scientists as well have attempted to discredit Rachel Carson's message, the greater portion of scientists and the general public are rallying behind the banner of conservation and the protection of our priceless natural resources. Men such as Paul R. Ehrlich, Barry Commoner, and Kenneth E. F. Watt among scientists, Senator Gaylord Nelson, Senator Edmund Muskie, and Representative Richard Ottinger among politicians, Arthur

Godfrey, Eddie Albert, and many other prominent lay persons have become effective spokesmen for the average American's concern for his environment.

But it began with pesticides. The first sign that something was amiss was the effect that the once-hailed insecticide DDT was having on the nation's birds of prey. Eagles and ospreys and some of the great hawks began to lay infertile eggs, or eggs whose membranes were too thin to contain their fragile cargo. DDT was traveling through the food chain, insect to fish to bird, accumulating in greater and greater densities at each link in the chain, finally impairing the bird's ability to reproduce. Should our birds of prey disappear, how will we control the numbers of mice and rodents in what remains of our fields and forests? What will be the effect on our crops and on the spread of disease carried by small animals? Where will it end?

DDT levels in the human food chain—in man himself—are on the upgrade. Mother's milk can truly be labeled unfit for human consumption. An enormous catch of salmon from Lake Michigan was confiscated a short while ago because of high DDT levels. Dairy products are endangered.

We do not yet know what the direct threat to mankind from DDT is, if any. But the chemical is accumulating in our bodies. Americans generally have enough of it in them to be classified as inedible under the standards set for foodstuffs. Although the threat is shadowy, the only wise course of action is to drop the use of this poison as fast as is possible. Several states have banned its use, and it seems clear the use of DDT in this country will cease within the next year or two.

But Latin American countries still use enormous (and unnecessary) quantities of DDT, and this tremendously persistent chemical does not break down and disappear after use. It is estimated that 2,000 tons of DDT are stored in the Antarctic snows and will be loosed upon the world in the future, no matter how soon the use of the chemical is completely stopped.

DDT alone is not to blame. Beautiful Lake Erie, of clear and pure water, is now dead. The children and adults who used to swim and play in the water and on the shore go there no longer. Those who fish in its once teeming waters catch old inner tubes, beer cans, and other products of our use-and-discard society. The stench of garbage and sewage permeates the breeze in summer, and the ice freezes grey in winter. Spring is truly silent there.

The white sands and crystal clear waters off Santa Barbara, California, are being permanently fouled by off-shore oil wells. Sea birds, their feathers coated with oil, lie dying in the gummy sludge at the edge of the sea. One of the most beautiful bodies of water in the world, brilliant Lake Tahoe in California, is slowly dying because developments around the lake use it as a cesspool.

"The public interest is rarely as ably represented legally in environmental issues as are private interests," Dr. Charles H. W. Foster told the first Conference on Law and Environment. The causes are not hard to pinpoint: lack of money for citizen and conservation groups to hire experienced ecologists and a scarcity of lawyers with experience in environmental cases even when funds have been available. Thus, while the conservation movement has never been healthier or more loaded with citizen-volunteers, it is at a decided disadvantage in the lengthy, time-consuming legal battle against encroachments on the environment from private and public sources.

But the evidence indicates that conservation and antipollution efforts are ideas whose time has come. Law schools are beginning to develop courses in environmental law. Books on the environment will help finance the efforts of the Sierra Club, a noted force in conservation. Recently the Senate voted one billion dollars to finance the fight against water pollution, to be used to help local communities and states finance waste treatment. Since 1958, the federal government has spent 1.2 billion and local government 4.2 billion dollars on fighting pollution. But our lakes and rivers and streams and even the ocean are getting worse, day after day.

We are at a pivotal point in the fight to save our planet. Will Man the polluter, become Man the protector? We do not know what we will become. But our hope lies in knowing what we are, and what we have been.

Cecil E. Johnson

THE
NATURAL
WORLD
Chaos and
Conservation

PART 1

INTRODUCTION

In 1962, a rebel of a book called "Silent Spring" was written by a quiet, exacting scientist named Rachel Carson. Since Miss Carson had always shunned publicity it must have taken considerable courage for her to write such a controversial book. For the first time a meaningful message had been written which forced scientists and nonscientists alike to take a long, lasting look at chemicals like DDT and their effects on our waters, forests, and fields and the plants and animals that make their homes there.

Since then, other spokesmen have emerged from the ranks of silent scientists and quiet citizens to speak out against man's rape of the world, and for the future of mankind on the planet. The battle lines have been drawn. The war against pollution has begun. "The dirty animal—man" has become aware of, become concerned about, and at last has begun to act on his suicidal tendencies toward the destruction of his planet.

Although the vested interests of commercialism (and some scientists as well) have attempted to discredit Rachel Carson's message, the greater proportion of scientists and the general public are rallying behind the banner of conservation and the protection of our priceless natural resources. Men such as Paul R. Ehrlich,

Barry Commoner, and Kenneth E. F. Watt, among scientists; Senator Gaylord Nelson, Senator Edmund Muskie, and Representative Richard Ottinger, among politicians; and Arthur Godfrey, Eddie Albert, and many other prominent lay persons have become effective spokesmen for the average American's concern for his environment.

It began with pesticides. The first sign that something was amiss was the effect that the once-hailed insecticide DDT was having on the nation's birds of prey. Eagles and ospreys and some of the great hawks began to lay infertile eggs, or eggs whose membranes were too thin to contain their fragile cargo. DDT was traveling through the food chain, from insect to fish to bird, accumulating in greater and greater densities at each link in the chain, finally impairing the bird's ability to reproduce. Should our birds of prey disappear, how will we control the numbers of mice and other rodents in what remains of our fields and forests? What will be the effect on our crops and on the spread of disease carried by small animals? Where will it end?

DDT levels in the human food—in man himself—are on the upgrade. Mother's milk can truly be labeled unfit for human consumption. An enormous catch of salmon from Lake Michigan was confiscated a short while ago because of high DDT levels. Cottage cheese, milk, and other dairy products are presently contaminated with undersirable levels of DDT.

We do not yet know what the direct threat to mankind from DDT is, but the chemical is accumulating in our bodies. Americans generally have enough of it in them to be classified as inedible under the standards set for foodstuffs. Although the threat is shadowy, the only wise course of action is to drop the use of this poison as fast as is possible. Several states have banned its use, and it seems clear that the use of DDT in this country will cease within the next few years.

Many Latin American countries still use enormous (and unnecesary) quantities of DDT, and this tremendously persistent chemical does not break down and disappear after use; instead it accumulates in the soils and waters of the world, later to be incorporated within the tissues of the plants and animals on which we rely for our food supply. It is estimated that 2,000 tons of DDT are stored in the Antarctic ice and will be released upon the world as the ice melts in the future, no matter how soon the use of the chemical is completed stopped.

The once beautiful Lake Erie, of clear and pure water, is now dead. The children and adults who used to swim in its waters and play on its shore go there no longer and are protesting loudly about the death of their beloved lake. Those who still fish in Lake Erie catch old inner tubes, beer cans, and other wastes of our use-and-discard society. The stench of garbage and sewage permeates the breeze in summer, and the ice freezes gray in the winter. Spring is truly silent there.

The white sands and crystal-clear waters off Santa Barbara, California, are being permanently fouled by offshore oil wells. Sea birds, their feathers coated with oil, lie dying in the gummy sludge at the edge of the sea. One of the most beautiful bodies of water in the world, brilliant Lake Tahoe in California, is slowly dying because housing developments around the lake use it as a cesspool.

"The public interest is rarely as ably represented legally in environmental issues as are private interests," Dr. Charles H. W. Foster told the first Conference on Law and Environment. The causes are not hard to pinpoint: lack of money for citizen and conservation groups to hire experienced ecologists, and a scarcity of lawyers with experience in environmental cases even when funds are available. Thus, while the conservation movement has never been healthier or more loaded with citizen-volunteers, it is at a decided disadvantage in the lengthy, time-consuming legal battle against encroachments on the environment from private and public sources.

But the evidence indicates that conservation and the antipollution efforts are ideas whose time has come. Law schools are beginning to develop courses in environmental law. Books on the environment will help finance the efforts of the Sierra Club, a noted force in conservation. Recently the Senate voted one billion dollars to support the fight against water pollution, to be used to help local communities and states finance the construction of waste treatment facilities. In the past 12 years, since 1958, the federal government has spent 1.2 billion and local government 4.2 billion dollars on fighting pollution. But our lakes and rivers and streams and even the ocean are getting worse, day after day.

We are at a pivotal point in the fight to save our planet. Will Man, the polluter, become Man, the protector? We do not know what we will become; but our hope lies in knowing what we are, and what we have been.

THE EARTH ANSWERS BACK *

WILLIAM VOGT

Dr. Vogt was one of the new breed of scientists who speak so eloquently in the areas of ecology and conservation; he desperately tried to alert the general public to the growing menace of man's destruction of the planet Earth. His now-famous book, "Road to Survival," has helped pioneer in the field of ecology. Dr. Vogt was Curator of Jones Beach State Bird Sanctuary, Chief of the Conservation Section of the Pan-American Union, and Director and Executive Vice-president of Planned Parenthood Association.

*From "Road to Survival" by William Vogt. Published by William Sloane Associates. Reprinted by permission of William Morrow & Company, Inc. Copyright © 1948 by Vogt.

he gray combers of the Tasman Sea raced past the ship from astern and with a slow rhythm, like that of a long pendulum, they lifted its screw to race futilely in the air. Captain Martins, after rising early as was his custom, stepped into the pilothouse to check the vessel's course, and then out onto the bridge. The low sun burned with a strange coppery yellow such as farther north might presage a typhoon. He looked anxiously about the horizon and estimated that the wind from the west was blowing with a force of five on the Beaufort scale. The sky did not look stormy, but he went once more into the pilothouse to check the glass. The barometric pressure was high, with no indication of trouble ahead.

Again he stepped out onto the bridge and scanned the sky. Then he became conscious that in the sheltered recesses of the bridge there was a thin overlay of brown dust. He muttered to himself in momentary irritation at the slackness of his crew, and looked again at the sky. He realized that, almost like a soft mist, a film of dust was falling through the air. In six hours he would be in Auckland. There was no land astern for more than a thousand miles. With a feeling of unbelief, he again checked the wind direction. His observation confirmed that the wind was coming from the west, and the fact that the dust must be blowing all the way from Australia.

He watched the sky a half hour before going in for his morning coffee. In Sydney he had heard talk of trouble in the back country, of hundreds of wells from which the windmills were sucking only air, of sheep dying of thirst or having to be slaughtered by hundreds of thousands. He had a cargo of wheat for England but this, too, was reported in short supply.

The soft dust was still raining down, and the reddening sun showed that the cloud was thickening. Martins shook his head and thought grimly to himself, "The whole damn continent must be blowing away."

MARIA

A few hours later, on the other side of the world in the Mexican State of Michoacán, a little woman trotted along a dusty road, with the swinging Indian gait that eats up the miles. On top of her head she balanced a rusty five-gallon tin that had not carried gasoline for many years. It was filled with her day's supply of water, precious water that had to be carried five miles to her pueblo. Ten miles every day she trotted, to have the liquid for tortillas and occasional tamales, for cooking her black beans and watering her few chickens. She weighed less than a hundred pounds and the can on her head was heavy, but the weight of her burden passed unnoticed in the heaviness of her heart. Until a few days ago she would also have been carrying a baby on her back, but now the rebozo was empty. As babies are so apt to do, in regions where water is scarce and polluted, hers had died early.

The ten-mile trip every day, under the beating sun of August or through the cold dry winds of January, did not seem in any way unusual to the woman. Didn't her husband have to walk even farther to till his little patch of maize and beans? She could neither read nor write, and she had no way of knowing that when her pueblo was built the people had gathered there because it was near a clear cold spring that gushed from the hillside. The sterile landscape about her, gray-stained with sparse grass and clumps of maguey, told her nothing of the rich forests that had once built soil for leagues about her town. She was tired, and her heart was heavy, but with the fatalism of a people that rarely knows surcease from a precarious existence, she sighed and muttered to herself, *"Se aguanta"*—one must bear with it. No phrase is more common on the lips of the women of her people.

TOM COBBETT

Tom Cobbett sat with his elbows on his desk and his face buried in his hands. He had come a long way since the first day he began to cut at the shining coal in the Yorkshire mines, but he was not thinking of the past, of the years on end when he had hacked out coal sixteen and eighteen hours a day nor of how he had forced himself to study night after night, pulling himself painfully up the ladder of books—the only escape he could find from the mines.

Tom should have been the happiest man in Great Britain tonight. He had just won a by-election with one of the greatest majorities piled up by any Labor candidate in England's history. His triumph was a tribute to himself and a rousing vote of support for the Labor party's program of socialization. For four hours he had put on an act for his friends and supporters, displayed a jubilation he did not feel. Far down in the honest recesses of his mind, Tom had to admit to himself that he wished it had not happened.

He was, he realized, a sort of way station in an historical process.

He had begun to fight for this election twenty years before, and he could not have avoided the victory without betraying his people and his country. But the weight of the prize was heavy on his mind. He, too, knew nothing of events in the Tasman Sea, nor on the dusty hillsides of Michoacán. If he had known of them, the orientation of his mind might have given him an inkling of their significance and increased the weight of the responsibility that lay so heavily upon him.

He was now a member of one of the oldest legislative bodies on the earth, charged for a time with participation in the ruling of a great empire. He and his party were committed to one of the most courageous and high-minded efforts man had ever made to better his lot.

Over two decades before the war, Cobbett had been fighting for this chance. Then the pattern he had been following was suddenly twisted out of shape by world revolution. Few members of the party saw this, but to Cobbett it was painfully clear. The Parliament to which he had just been elected ruled fifty million people who lived on two islands with an area of 95,000 square miles, the size of Oregon. By the most heroic efforts of men and women together, utilizing every yard of available land, these people had not been able to produce much more than half of the food they needed.

Before World War I this had not mattered—much. The coal he had hewn from the dripping seams had been used to buy the beef that meant so much to England, and the corn, as the English called it, for their bread. The remarkable skills of British workers had fashioned the raw products of all the world into a multitude of things that five continents would purchase with food.

But now the mines were playing out and British skills had been duplicated, with varying degrees of success, around the world. The British worker's horizon had broadened, and he looked to the prosperity of the American worker and the economic organization of the Russian, and claimed them for his own. Cobbett, along with his party, had taken on the responsibility of satisfying this claim. Indeed he felt an intuitive kinship with the Australia of the blowing earth, with the Argentina of the rich lands, with the North America of failing waters. The hungers and desires of his crowded island would somehow have to be meshed with the rest of the world but now, faced with a driving reality, Cobbett knew that a political and economic solution would not be enough.

FOSTER RAMSEY

From far above, in the deep-blue sky, came the vagrant trumpeting of homing geese. There was not another sound in the still prairie night, and Foster Ramsey put down his pen and listened. As always, he felt a tug at his heart when the geese passed over. Across the high plains, the black spruce, the tamarack bogs, and the tundra, the flock would drive northward until it split over isolated pools, to set up the season's housekeeping.

He took off his glasses, rubbed his tired eyes, and shook his head ruefully. Wild geese to income taxes—it was a long jump! "Not a chance," he said to himself. "The twins will have to make do here."

He was president of a college but it was a small college, a "cow college." His salary would not have impressed an automobile salesman in New York. If there had been only the two older children to consider, things would have been much simpler. The twins had been a surprise, and the extra burden they brought with them had been enormously complicated by the war. The high school in the little town where they lived was not much good. How could it be, when the teachers received only $1,500 a year? Ramsey had hoped the children might go away to boarding school, but even with the scholarships he could count on, he knew he could not swing it.

He looked down again at the tax form, and wryly signed his name. Then he wrote a check. Taxes this year, including the hidden ones, would take nearly a third of his income. "Seventeen weeks out of the year," he said to himself, "I worked for the government. Thank God, it's the American government, and not the Nazi government!" In a sense he was grateful to be paying these taxes, glad he was able to, but as he looked at the form in front of him he could not help thinking what pleasure it, and millions like it, would have given the defeated Nazi hierarchy.

Germany had organized the greatest system of slave labor the world had ever known. Few people, Ramsey knew, realized that this slavery was not yet at an end. The major part of his taxes was helping to pay for the war and its aftermath. With luck he would have thirty years more of work ahead of him. If the national debt were ever to be paid, if there were not to be repudiation or disastrous inflation, he knew there could not be much hope that his tax load would be reduced. Seventeen weeks a year for thirty years—ten years of his life dedicated primarily to paying for the adventures of Mussolini, Hitler, and the Japanese war lords. Multiply his lot by that of tens of millions of other American workers, and it added up to a greater force of slave laborers than had ever struggled under the Nazi lash in Europe.

Ramsey did not know of the sea captain, the unhappy Tarascan woman, the English politician, but if he had, his well-organized mind, used to thinking in terms of the land, would have readily related them to the papers on his desk. Because he understood the land, had watched the degeneration of his state's cattle ranges, had seen failing springs break more than one rancher, he would have recognized that a gullied hillside in Szechwan or a hollow-eyed miner in the Ruhr was a factor, although a hidden factor, in his tax form.

He had already recognized the substantial decrease in his living standard. He knew that for the rest of his life he would not only be carrying his share of the burden of the war but contributing increasingly to what he mentally called the "gimme boys." Some of them wore union badges, some of them carried cards of the American Farm Bureau, some of them wore the little blue cap of the American Legion to which service in World War I entitled him. He might have added to them worried government officials in capitals in many foreign countries.

No, the kids wouldn't get much of a high school education. He and Janet would have to see what they could do to fill in the gaps at home.

JIM HANRAHAN

Jim Hanrahan stood in front of the mirror in his bathroom door. He tugged speculatively at the tire around his middle. It was many a year since he had pulled a peavey and he knew he looked it; but he shrugged his shoulders and thought to himself, "What the hell!" From now on peaveys would be handled for him by other men, and they would get one-fiftieth of what he made from their work.

He went to the window and looked across at the magnificent spectacle of the snow-capped mountains, unusually clear this afternoon. He poured himself a stiff highball and sighed contentedly. In a couple of hours he would eat the best dinner to be had in the city, and after that there would be the little girl at the Waikiki night club.

Tonight was a celebration, and he was going to make the most of it. He had cleaned up on war contracts—working eighteen hours a day—and now he had cashed in. True, he had had to pay a politico $8,000, but even so he had been able to get timber covering five entire mountains for less than a fifth of what it would have cost him in the States. His plant was well capitalized and his engineers would be on the ground within two weeks.

In cold dusk the night before he had stood among the mighty trees and looked out over another city below. He was buying the city's water supply, and he knew it. But he had a lumberman's conscience, with calluses so thick that he no longer felt even cynical about these deals. It was their country. If they wanted to sell it, it was O.K. with him. This one job would leave him sitting pretty for at least fifteen or twenty years. What would happen to the town without water at the end of that period he did not even consider. After all, business is business!

THE NAMELESS

The ship was running without lights. The skipper would be damned glad to get rid of his cargo; but even so he kept the screw turning at only quarter speed. On the decks, small groups of huddled forms clung together and prayed. They knew that before morning some of them would probably be dead.

Death was nothing new to this human cargo. Most of them had lived with it as a familiar for ten years or more. Battered about from town to town, from country to country, they had wandered in an almost unbroken wilderness of dislike, of distress and stupid hatred. Not one of the men, women and children aboard but had lost someone close and dear to him. Some of these had died in concentration camps, others had been shot in cold blood as hostages, still others had been ripped apart by the whistling bombs. Hundreds of miles these people had walked, and many had left bloody tracks behind them on stony roads and in the snows. For years they had roamed, almost without hope, but now they were coming out of the wilderness—truly into a promised land.

For most of their lives, it seemed, they had lived as human contraband, and they were still human contraband. Armed guards lay in wait for them but some of them, perhaps most of them, would be able to slip ashore in the darkness. The guards might well shoot them down, although they themselves had no reason to feel any hatred against these tired wanderers. The guards themselves were mere puppets manipulated by a politician's mistake. Lies had been spoken twenty-five years before, and these guards were set, somehow to turn the lies into truth.

The promised land to which every man and woman and child in this group hoped to win through was no land of milk and honey, such as their ancestors had found. It was a worn-out desert that once had been a rich landscape, that once had supported towns and industries and sent its ships to the edge of the known west. Man's abuse had wrung most of the life out of it, and now man's intelligence and back-breaking toil were slowly bringing it to life again. Grueling labor, harsh rations, little rest, were the best these few score people could hope for if they managed to scuttle past the whistling rifle bullets. But in the heart of every one of them there was high hope and confidence. They knew what their people had done and they knew what they could do. Here, although they might eventually have to fight for it, they were sure they could work in peace.

To these pitiful wanderers through the storms of a hostile world there was no word more precious than "peace."

WONG

Wong sat by the side of the dusty road, almost too weak to hold himself upright. Carts with shouting drivers, wheelbarrows piled high and pushed by silent men, occasionally rickshaws, passed by, often within a few inches. It seemed as though they must strike him, but he did not move.

Wong no longer cared. He knew he was going to die, because he had seen hundreds dying all about him. He felt no pain. This had passed days ago. He no longer even felt hungry, and this was such a new experience to him that it almost made him welcome his approaching death. He looked sixty and was thirty-four. Hardly a day since he had been weaned had he known what it was to escape the hunger pains that gnawed at his middle. His bones stuck out through a yellow skin even more parchment-like than usual.

Three weeks before, or four, or five, he did not remember, he had left his wife and baby in the western provinces and started to walk toward the sea, in the hope of finding something for them to eat. So far as he knew, he did not have a friend or a relative within hundreds of miles. One by one he had sold the garments off his back, until he was all but naked. Once in a while he had been able to scrape together a few grains of rice, but for days now he had known that he was losing the struggle. There had not been enough to keep him going, and in that great world of famine where hundreds and thousands and tens of thousands were dying as he was, no one cared. To a European or an American this might have seemed the ultimate sadness. To Wong it was meaningless, since he was dying as he had lived most of his life.

The sun seemed like an incandescent dome clamped atop the world. Not even in the great river bed was there a trickle of water. Eight months ago it had raged with floods that broke into the towns and stormed away with hundreds of victims. Now he would have to sell his last garment even to get a drink. The steep, bare hillsides, completely denuded of vegetation, towered above him. The cold wind flung towers of dust before it, dust that powdered into the corners of his mouth.

Halfway around the world men and women were trying to scrape together such grain as they could, to keep the ember of life glowing in Wong and millions like him. But he knew nothing of this. If he had, it would have made no difference. He knew now, when he had finally given up all hope of life, that nowhere in the world could there be enough food to feed so many hungry mouths.

JOE SPENCER

Joe Spencer's hand trembled as he moved the slide back and forth under the microscope's objective. Over and over again he checked the blood smear. There was no trace of plasmodium.

He sat back in his chair and let his eyes wander about the laboratory. The monkeys and the birds played in their cages, quite unconcerned at having a date with history. All through the war Joe had worked, as had hundreds of other researchers, to find a certain and harmless protection against malaria for United Nations troops. Now with the war, technically, at least, over, he apparently had it.

He had had 100 per cent success in protecting the experimental animals with a few grains of white powder. Over and over again he had run tests on them, and all had been negative. Then, after dosing with the compound, he had deliberately inoculated himself. It was not at all certain that this chemical, whose molecules he had been shuffling for nearly two years, would work the same way on him as it had on canaries. He had waited two weeks for a positive reaction—and there had been none. He had tried again, with negative results. Three more checks indicated that he was completely immune. Then Marion, his wife, insisted that he try it on her before saying anything to anyone else about it. A slide smeared with his wife's blood lay in front of him. If this was negative, he would be 99 per cent certain. He slipped the slide onto the stage of the microscope and with practiced fingers ran over it rapidly. Not a thing to be seen. Then he settled down and checked it with infinite care, by minute fractions of a millimeter. Still nothing!

He sat back in his chair, and realized that he had broken into a heavy sweat. Unless he was most improbably wrong, he had a sure and harmless preventive for one of the worst man killers of all time. Its manufacture would cost less than aspirin, and millions of suffering men and women would find immediate surcease from racking pains. What would this mean to the hordes of India and China? What would it mean to the world?

Joe knew that in the test tubes of his laboratory there was confined a power that was perhaps as dangerous as that of the atomic bomb. He had walked the cobblestone streets of Rumanian towns, of muddy byways in Italy, and seen hundreds of coffins carried on the shoulders of men, coffins that had been filled by the bites of mosquitoes. From afar off he had seen the burning ghats of India, had watched them across stinking pools of standing water from which the winged death had silently flitted away. His fingers had searched for the swollen spleens of children in Guayaquil and Manáos, and all over northern South America. He had watched the dragging gait of men and women who spent their nights burning with the fire they called "paludismo."

Then he had spent two weeks in Puerto Rico, where the miracles of American medicine had been worked, with the chief result that more people were kept alive to live more miserably. He knew India had grown by some fifty millions in ten years, and even before the first of those fifty millions was born there was not enough food to go around. Was there any kindness in keeping people from dying of malaria so that they could die more slowly of starvation? Could there be any end of wars, and rumors of wars, while such people as the prosperous Americans had far more than they needed and the millions of India and China and Java and Western Europe—and perhaps Russia—did not have enough? Few men of his day, Joe knew, had had the power to shake so profoundly the future of the world.

None of these human beings, sharing similar hopes and despair according to his individual lot, is aware of the others. None of them, except the scientist, sees himself as part of a great world drama in which each plays his part as both cause and effect. This gallery of portraits of people today, of the troubled or contented, could be multiplied millions of times. They are not all of humanity, but they are a sample of all of humanity. They are fictional, but it would not be difficult to find living equivalents.

All of them have one thing in common. The lot of each, from Australian sea captain to biochemist, is completely dependent on his or her global environment, and each one of them in greater or less degree influences that environment. One common denominator controls their lives: the ratio between human populations and the supply of natural resources, with which they live, such as soil, water, plants, and animals. This is a highly unstable relationship, changing from moment to moment, continually conditioned anew by human acts.

Before the great age of exploration at the end of the fifteenth century, this relationship was a simple matter. Man lived in a series of isolated cells. What was done in Britain had little influence on what was done in China. What was done in the great Mississippi Basin made no impact on the rest of the world. Then Columbus set in motion forces that only a few people have yet begun to comprehend.

Columbus, more than the atomic scientists, made this one geographic world. Woodrow Wilson saw that we all live in one world in a political sense, and Wendell Willkie popularized the concept for the man in the street. However, few of our leaders have begun to understand that we live in one world in an ecological—an environmental—sense. Dust storms in Australia have an inescapable effect on the American people; they set mutton prices soaring, and our own western sheep raisers, not looking beyond their own depleted ranges to the starving peoples of El Salvador and Greece, are content.

The little Indian woman, whose forebears were robbed by the cupidity of early Spanish miners, is a drop in the great pool of deprivation that is threatening to overwhelm the world. She is an integral part of our own Latin-American relations; whether or not she and her twenty-three million compatriots have enough to eat and drink and wear is rich in meaning for the foreign policy of the United States.

The English politician haunted by the chimera of high American living standards, is, though he may not see it, the resultant of forces that stirred into life the day the "Santa Maria's" lookout cried "Land!" to his mutinous fellow sailors.

The American lumberman, riding high, wide and handsome, neither knows nor cares that he is stretching the slow fingers of death throughout the New World; he is a true worshiper of that sacred cow Free Enterprise.

The refugees, who have suffered perhaps more greatly than any other group of people in history, give no thought to the fact that their fate has its roots deep in the cotton lands of the Mississippi Delta, the black loam of Iowa cornfields, the forests of the Appalachians. They only see, with hope high in their hearts, that they are returning to rebuild a land devastated by their ancestors some three thousand years ago.

Wong, dying by the side of the road, cannot possibly comprehend that he is being killed by an unexpected genie that spiraled out of the test tubes and cultures of Louis Pasteur. Nor does he see himself as part of an explosive pressure rapidly building and threatening someday to burst all bonds.

The biologist sees his role more clearly. Engaged in a mission of mercy, he realizes that his little white grains of powder may turn into dragon's teeth. He hesitates as to whether he should loose this new, uncontrolled force upon the world. If he is like most scientists, he is not politically minded and has little idea about ways and means of guiding the forces of science.

Political leaders are trying to solve the infinitely complex problems of the modern world while literally ignoring most of that world. They assume, perhaps because most of them are city men and therefore urban minded, that man lives in a vacuum, independent of his physical environment. They are seeking the solution of an extraordinarily complex equation and almost completely neglecting major factors.

Perhaps a bio-equation, that takes into account man's physical universe, will help us to clear our thinking and even to regulate the forces that bemuse our political leaders. Here is a simple formula, no more complicated than the relationship of the family budget to the family income.

This formula is $C = B:E$.

Here C stands for the *carrying capacity* of any area of land. In its simplest form this means its ability to provide food, drink, and shelter to the creatures that live on it. In the case of human beings, the equation finds complicated expression in terms of civilized existence.

B means *biotic potential*, or the ability of the land to produce plants for shelter, for clothing, and especially for food. Only plants are able to synthesize food from the raw materials of the earth and the air in a form assimilable by animals. This is the only way in which food exists for animals.

E stands for *environmental resistance*, or the limitations that any environment, including the part of it contrived and complicated by man, places on the biotic potential or productive ability. *The carrying capacity is the resultant of the ratio between the other two factors.*

The equation is, perhaps, oversimplified, but it expresses certain relationships—almost universally ignored—that every minute of every day touch the life of every man, woman, and child on the face of the globe.

Until an understanding of these relationships on a world scale enters into the thinking of free men everywhere, and into the thinking of rulers of men who are not free, there is no possibility of any considerable improvement of the lot of the human race. Indeed, if we continue to ignore these relationships, there is little probability that mankind can long escape the searing downpour of war's death from the skies.

And when this comes, in the judgment of some of the best informed authorities, it is probable that at least three-quarters of the human race will be wiped out.

POLLUTING THE ENVIRONMENT *

LORD RITCHIE-CALDER

Lord Ritchie-Calder, a prominent newspaperman and author, has been concerned with ecological problems. In this essay, *Polluting the Environment*, he speaks in eloquent fashion in words whose meanings cannot be mistaken.

He is the author of "The Inheritors," "Living with the Atom," "Common Sense about a Starving World," and "Man and the Cosmos."

* Reprinted, by permission, from the May 1969 issue of *The Center Magazine*, a publication of the Center for the Study of Democratic Institutions in Santa Barbara, California.

To hell with posterity! After all, what have the unborn ever done for us? Nothing. Did they, with sweat and misery, make the Industrial Revolution possible? Did they go down into the carboniferous forests of millions of years ago to bring up coal to make wealth and see nine-tenths of the carbon belched out as chimney soot? Did they drive the plows that broke the plains to release the dust that the buffalo had trampled and fertilized for centuries? Did they have to broil in steel plants to make the machines and see the pickling acids poured into the sweet waters of rivers and lakes? Did they have to labor to cut down the tall timbers to make homesteads and provide newsprint for the Sunday comics and the celluloid for Hollywood spectaculars, leaving the hills naked to the eroding rains and winds? Did they have the ingenuity to drill down into the Paleozoic seas to bring up the oil to feed the internal-combustion engines so that their exhausts could create smog? Did they have the guts to man rigs out at sea so that boreholes could probe for oil in the offshore fissures of the San Andreas Fault? Did they endure the agony and the odium of the atom bomb and spray the biosphere with radioactive fallout? All that the people yet unborn have done is to wait and let us make the mistakes. To hell with posterity! That, too, can be arranged. As Shelley wrote: "Hell is a city much like London, a populous and smoky city."

At a conference held at Princeton, New Jersey, at the end of 1968, Professor Kingsley Davis, one of the greatest authorities on urban development, took the role of hell's realtor. The prospectus he offered from his latest survey of world cities was hair-raising. He showed that thirty-eight per cent of the world's population is already living in what are defined as "urban places." Over one-fifth of the world's population is living in cities of a hundred thousand or more. Over 375,000,000 people are living in cities of a million and over. On present trends it will take only fifteen years for half the world's population to be living in cities, and in fifty-five years everyone will be urbanized.

Davis foresaw that within the lifetime of a child born today, on present rates of population increase, there will be fifteen billion people to be fed and housed—over four times as many as now. The whole human species will be living in cities of a million and over and the biggest city will have 1,300,000,000 inhabitants. Yes, 1.3 billion. That is 186 times as many as there are in Greater London today.

In his forebodings of Dystopia (with a "y" as in dyspepsia, but it could just as properly be "Dis," after the ruler of the Underworld),

Doxiades has warned about the disorderly growth of cities, oozing into each other like confluent ulcers. He has given us Ecumenopolis—World City. The East Side of Ecumenopolis would have as its Main Street the Eurasian Highway, stretching from Glasgow to Bangkok, with the Channel tunnel as an underpass and a built-up area all the way. West Side, divided not by railroad tracks but by the Atlantic, is already emerging (or, rather, merging) in the United States. There is talk, and evidence, of "Boswash," the urban development of a built-up area from Boston to Washington. On the Pacific Coast, with Los Angeles already sprawling into the desert, the realtor's garden cities, briskly reenforced by industrial estates, are slurring into one another and presently will stretch all the way from San Diego to San Francisco. The Main Street of Sansan will be Route 101. This is insansanity. We do not need a crystal ball to foresee what Davis and Doxiades are predicting—we can see it through smog-colored spectacles; we can smell it seventy years away because it is in our nostrils today; a blind man can see what is coming.

Are these trends inevitable? They are unless we do something about them. I have given up predicting and have taken to prognosis. There is a very important difference. Prediction is based on the projection of trends. Experts plan for the trends and thus confirm them. They regard warnings as instructions. For example, while I was lecturing in that horror city of Calcutta, where three-quarters of the population live in shacks without running water or sewage disposal, and, in the monsoon season, wade through their own floating excrement, I warned that within twenty-five years there would be in India at least five cities, each with populations of over sixty million, ten times bigger than Calcutta. I was warning against the drift into the great conurbations now going on, which has been encouraged by ill-conceived policies of industrialization. I was warning against imitating the German Ruhr, the British Black Country, and America's Pittsburgh. I was arguing for "population dams," for decentralized development based on the villages, which make up the traditional cultural and social pattern of India. These "dams" would prevent the flash floods of population into overpopulated areas. I was *warning*, but they accepted the prediction and ignored the warning. Soon thereafter I learned that an American university had been given a contract to make a feasibility study for a city of sixty million people north of Bombay. When enthusiasts get busy on a feasibility study, they invariably find that it is feasible. When they get to their drawing boards they have a whale of a time. They design skyscrapers above ground and subterranean tenements below ground. They work out minimal requirements of air and hence how much breathing space a family can survive in. They design "living-units," hutches for battery-fed people who are stacked together like kindergarten blocks. They provide water and regulate the sewage on the now well-established cost-efficiency principles of factory-farming. And then they finish up convinced that this is the most economical way of housing people. I thought I had scotched the idea by making representations through influential Indian friends. I asked them, among other things, how many mental hospitals they were planning to take care of the millions who would surely go mad under such conditions. But I have heard rumors that the planners are so slide-rule happy they are planning a city for six hundred million.

Prognosis is something else again. An intelligent doctor, having diagnosed the symptoms and examined the patient's condition, does not say (except in soap operas): "You have six months to live." He says: "Frankly, your condition is serious. Unless you do so-and-so, and unless I do so-and-so, it is bound to deteriorate." The operative phrase is "do so-and-so." One does not have to plan *for* trends; if they are socially undesirable our duty is to plan *away* from them, and treat the symptoms before they become malignant.

A multiplying population multiplies the problems. The prospect of a world of fifteen billion people is intimidating. Three-quarters of the world's present population is inadequately fed—hundreds of millions are not getting the food necessary for well-being. So it is not just a question of quadrupling the present food supply; it means six to eight times that to take care of present deficiencies. It is not a matter of numbers, either; it is the *rate* of increase that mops up any improvements. Nor is it just a question of housing but of clothing and material satisfactions—automobiles, televisions, and the rest. That means greater inroads on natural resources, the steady destruction of amenities, and the conflict of interest between those who want oil and those who want oil-free beaches, or between those who want to get from here to there on wider and wider roads and those whose homes are going to collapse in mud slides because of the making of those roads. Lewis Mumford has suggested that civilization really began with the making of containers—cans, non-returnable bottles, cartons, plastic bags, none of which can be redigested by nature. Every sneeze accounts for a personal tissue. Multiply that by fifteen billion.

Environmental pollution is partly rapacity and partly a conflict of interest between the individual, multimillions of individuals, and the commonweal; but largely, in our generation, it is the exaggerated effects of specialization with no sense of ecology, i.e. the balance of nature. Claude Bernard, the French physiologist, admonished his colleagues over a century ago: "True science teaches us to doubt and in ignorance to refrain." Ecologists feel their way with a detector through a minefield of doubts. Specialists, cocksure of their own facts, push ahead, regardless of others.

Behind the sky-high fences of military secrecy, the physicists produced the atomic bomb—just a bigger explosion—without taking into account the biological effects of radiation. Prime Minister Attlee, who consented to the dropping of the bomb on Hiroshima, later said that no one, not Churchill, nor members of the British Cabinet, nor he himself, knew of the possible genetic effects of the blast. "If the scientists knew, they never told us." Twenty years before, Hermann Muller had shown the genetic effects of radiation and had been awarded the Nobel Prize, but he was a biologist and security treated this weapon as a physicist's bomb. In the peacetime bomb-testing, when everyone was alerted to the biological risks, we were told that the fallout of radioactive materials could be localized in the testing grounds. The radioactive dust on The Lucky Dragon, which was fishing well beyond the proscribed area, disproved that. Nevertheless, when it was decided to explode the H-bomb the assurance about localization was blandly repeated. The H-bomb would punch a hole into the stratosphere and the radioactive gases would dissipate. One of those gases is radioactive krypton, which decays into radioactive strontium, a particulate. Somebody must have known that but nobody worried unduly because it would happen above the troposphere, which might be described as the roof of the weather system. What was definitely overlooked was the fact that the troposphere is not continuous. There is the equatorial troposphere and the polar troposphere and they overlap. The radioactive strontium came back through the transom and was spread all over the world by the climatic jet streams to be deposited as rain. The result is that there is radiostrontium (which did not exist in nature) in the bones of every young person who was growing up during the bomb-testing—every young person, everywhere in the world. It may be medically insignificant but it is the brandmark of the Atomic Age generation and a reminder of the mistakes of their elders.

When the mad professor of fiction blows up his laboratory and then himself, that's O.K., but when scientists and decision-makers act out of ignorance and pretend it is knowledge, they are using the biosphere, the living space, as an experimental laboratory. The whole world is put in hazard. And they do it even when they are told not to. During the International Geophysical Year, the Van Allen Belt was discovered. The Van Allen Belt is a region of magnetic phenomena. Immediately the bright boys decided to carry out an experiment and explode a hydrogen bomb in the Belt to see if they could produce an artificial aurora. The colorful draperies, the luminous skirts of the aurora, are caused by drawing cosmic particles magnetically through the rare gases of the upper atmosphere. It is called ionization and is like passing electrons through the vacuum tubes of our familiar neon lighting. It was called the Rainbow Bomb. Every responsible scientist in cosmology, radio-astronomy, and physics of the atmosphere protested against this tampering with a system we did not understand. They exploded their bomb. They got their pyrotechnics. We still do not know the price we may have to pay for this artificial magnetic disturbance.

We could blame the freakish weather on the Rainbow Bomb but, in our ignorance, we could not sustain the indictment. Anyway, there are so many other things happening that could be responsible. We can look with misgiving on the tracks in the sky—the white tails of the jet aircraft and the exhausts of space rockets. These are introducing into the climatic system new factors, the effects of which are immensurable. The triggering of rain clouds depends upon the water vapor having a toehold, a nucleus, on which to form. That is how artificial precipitation, so-called rainmaking, is produced. So the jets, crisscrossing the weather system, playing tic-tac-toe, can produce a man-made change of climate.

On the longer term, we can see even more drastic effects from the many activities of *Homo insapiens*, Unthinking Man. In 1963, at the United Nations Science and Technology Conference, we took stock of the several effects of industrialization on the total environment.

The atmosphere is not only the air which humans, animals, and plants breathe; it is the envelope which protects living things from harmful radiation from the sun and outer space. It is also the medium of climate, the winds and the rain. These are inseparable from the hydrosphere, including the oceans, which cover seven-tenths of the earth's surface with their currents and evaporation; and from the biosphere, with the vegetation and its transpiration and photosynthesis; and from the lithosphere, with its minerals, extracted for man's increasing needs. Millions of years ago the sun encouraged the growth of the primeval forests, which became our coal, and the life-growth in the Paleozoic seas, which became our oil. Those fossil-fuels, locked in the vaults through eons of time, are brought out by modern man and put back into the atmosphere from the chimney stacks and exhaust pipes of modern engineering.

This is an overplus on the natural carbon. About six billion tons of primeval carbon are mixed with the atmosphere every year. During the past century, in the process of industrialization, with its burning of fossil-fuels, more than four hundred billion tons of carbon have been artificially introduced into the atmosphere. The concentration in the air we breathe has been increased by approximately ten per cent; if all the known reserves of coal and oil were burned the concentration would be ten times greater.

This is something more than a public-health problem, more than a question of what goes into the lungs of the individual, more than a question of smog. The carbon cycle in nature is a self-adjusting mechanism. One school of scientific thought stresses that carbon monoxide can reduce solar radiation. Another school points out that an increase in carbon dioxide raises the temperature at the earth's surface. They are both right. Carbon dioxide, of course, is indispensable for plants and hence for the food cycle of creatures, including humans. It is the source of life. But a balance is maintained by excess carbon being absorbed by the seas. The excess is now taxing this absorption, and the effect on the heat balance of the earth can be significant because of what is known as "the greenhouse effect." A greenhouse lets in the sun's rays and retains the heat. Similarly, carbon dioxide, as a transparent diffusion, does likewise; it admits the radiant heat and keeps the convection heat close to the surface. It has been estimated that at the present rate of increase (those six billion tons a year) the mean annual temperature all over the world might increase by $5.8°F$. in the next forty to fifty years.

Experts may argue about the time factor or about the effects, but certain things are observable not only in the industrialized Northern

Hemisphere but also in the Southern Hemisphere. The ice of the north polar seas is thinning and shrinking. The seas, with their blanket of carbon dioxide, are changing their temperatures with the result that marine life is increasing and transpiring more carbon dioxide. With this combination, fish are migrating, even changing their latitudes. On land, glaciers are melting and the snow line is retreating. In Scandinavia, land which was perennially under snow and ice is thawing. Arrowheads of a thousand years ago, when the black earth was last exposed and when Eric the Red's Greenland was probably still green, have been found there. In the North American sub-Arctic a similar process is observable. Black earth has been exposed and retains the summer heat longer so that each year the effect moves farther north. The melting of the sea ice will not affect the sea level because the volume of floating ice is the same as the water it displaces, but the melting of the land's ice caps and glaciers, in which water is locked up, will introduce additional water to the oceans and raise the sea level. Rivers originating in glaciers and permanent snowfields (in the Himalayas, for instance) will increase their flow, and if the ice dams break the effects could be catastrophic. In this process, the patterns of rainfall will change, with increased precipitation in areas now arid and aridity in places now fertile. I am advising all my friends not to take ninety-nine-year leases on properties at present sea level.

The pollution of sweet-water lakes and rivers has increased so during the past twenty-five years that a Freedom from Thirst campaign is becoming as necessary as a Freedom from Hunger campaign. Again it is a conflict of motives and a conspiracy of ignorance. We can look at the obvious—the unprocessed urban sewage and the influx of industrial effluents. No one could possibly have believed that the great Lakes in their immensity could ever be overwhelmed, or that Niagara Falls could lose its pristine clearness and fume like brown smoke, or that Lake Erie could become a cesspool. It did its best to oxidize the wastes from the steel plants by giving up its free oxygen until at last it surrendered and the anaerobic microörganisms took over. Of course, one can say that the mortuary smells of Lake Erie are not due to the pickling acids but to the dead fish.

The conflict of interests amounts to a dilemma. To insure that people shall be fed we apply our ingenuity in the form of artificial fertilizers, herbicides, pesticides, and insecticides. The runoff from the lands gets into the streams and rivers and distant oceans. DDT from the rivers of the United States has been found in the fauna of the Antarctic, where no DDT has ever been allowed. The dilemma becomes agonizing in places like India, with its hungry millions. It is now believed that the new strains of Mexican grain and I.R.C. (International Rice Center in the Philippines) rice, with their high yields, will provide enough food for them, belly-filling if not nutritionally balanced. These strains, however, need plenty of water, constant irrigation, plenty of fertilizers to sustain the yields, and tons of pesticides because standardized pedigree plants are highly vulnerable to disease. This means that the production will be concentrated in the river systems, like the Gangeatic Plains, and the chemicals will drain into the rivers.

The glib answer to this sort of thing is "atomic energy." If there is enough energy and it is cheap enough, you can afford to turn rivers into sewers and lakes into cesspools. You can desalinate the seas. But, for the foreseeable future, that energy will come from atomic fission, from the breaking down of the nucleus. The alternative, promised but undelivered, is thermonuclear energy—putting the H-bomb into dungarees by controlling the fusion of hydrogen. Fusion does not produce waste products, fission does. And the more peaceful atomic reactors there are, the more radioactive waste there will be to dispose of. The really dangerous material has to be buried. The biggest disposal area in the world is at Hanford, Washington. It encloses a stretch of the Columbia River and a tract of country covering 650 square miles. There, a twentieth-century Giza, it has cost much more to bury live atoms than it cost to entomb all the mummies of all the Pyramid Kings of Egypt.

At Hanford, the live atoms are kept in tanks constructed of carbon steel, resting in a steel saucer to catch any leakage. These are enclosed in a reënforced concrete structure and the whole construction is buried in the ground with only the vents showing. In the steel sepulchers, each with a million-gallon capacity, the atoms are very much alive. Their radioactivity keeps the acids in the witches' brew boiling. In the bottom of the tanks the temperature is well above the boiling point of water. There has to be a cooling system, therefore, and it must be continuously maintained. In addition, the vapors generated in the tanks have to be condensed and scrubbed, otherwise a radioactive miasma would escape from the vents. Some of the elements in those high-level wastes will remain radioactive for at least 250,000 years. It is most unlikely that the tanks will endure as long as the Egyptian pyramids.

Radioactive wastes from atomic processing stations have to be transported to such burial grounds. By the year 2000, if the present practices continue, the number of six-ton tankers in transit at any given time would be well over three thousand and the amount of radioactive products in them would be 980,000,000 curies—that is a mighty number of curies to be roaming around in a populated country.

There are other ways of disposing of radioactive waste and there are safeguards against the hazards, but those safeguards have to be enforced and constant vigilance maintained. There are already those who say that the safety precautions in the atomic industry are excessive.

Polluting the environment has been sufficiently dramatized by events in recent years to show the price we have to pay for our reckless-ness. It is not just the destruction of natural beauty or the sacrifice of recreational amenities, which are crimes in themselves, but interfer-ence with the whole ecology—with the balance of nature on which persistence of life on this planet depends. We are so fascinated by the gimmicks and gadgetry of science and technology and are in such a hurry to exploit them that we do not count the consequences.

We have plenty of scientific knowledge but knowledge is not wis-dom: wisdom is knowledge tempered by judgment. At the moment, the scientists, technologists, and industrialists are the judge and jury in their own assize. Statesmen, politicians, and administrators are ill-equipped to make judgments about the true values of discoveries or developments. On the contrary, they tend to encourage the crash pro-grams to get quick answers—like the Manhattan Project, which turned the laboratory discovery of uranium fission into a cataclysmic bomb in six years; the Computer/Automation Revolution; the Space Pro-gram; and now the Bio-engineering Revolution, with its possibilities not only of spare-organ plumbing but of changing the nature of living things by gene manipulation. They blunder into a minefield of unde-tected ignorance, masquerading as science.

The present younger generation has an unhappy awareness of such matters. They were born into the Atomic Age, programmed into the Computer Age, rocketed into the Space Age, and are poised on the threshold of the Bio-engineering Age. They take all these marvels for granted, but they are also aware that the advances have reduced the world to a neighborhood and that we are all involved one with another in the risks as well as the opportunities. They see the mistakes writ large. They see their elders mucking about with *their* world and *their* future. That accounts for their profound unease, whatever forms their complaints may take. They are the spokesmen for posterity and are justified in their protest. But they do not have the explicit answers, either.

Somehow science and technology must conform to some kind of social responsibility. Together, they form the social and economic dynamic of our times. They are the pacesetters for politics and it is in the political frame of reference that answers must be found. There can never be any question of restraining or repressing natural curiosity, which is true science, but there is ample justification for evaluating and judging developmental science. The common good requires nothing less.

THE SCIENTIST AND THE CITIZEN *

BARRY COMMONER

Barry Commoner is Professor of Plant Physiology and Chairman of the Department of Botany at Washington University. For several years he was Chairman of the Committee on Science in the Promotion of Human Welfare of the American Association for the Advancement of Science.

In his hard-hitting, best-selling "Science and Survival," Professor Commoner steps onto the stage of controversy. His is an eloquent plea to every citizen to join in the task of safeguarding the future of mankind.

The New Yorker magazine said, "Professor Commoner belongs to a [special] species of scientist—those who are committed to increased public understanding of the social and political consequences of science."

his account has thus far been limited to matters of science and technology, and most of what I have said can be supported or disputed by marshaling appropriate data, either scientific or historical. Thus far, however, I have described problems but no solutions. Since it is impossible to speak of what might be done to alleviate environmental pollution or to save us from the frightful catastrophe of nuclear war without rendering opinions and expressing personal convictions, we must now move from the realm of science and technology into these less certain grounds.

That we *have* learned a little about how to cope with some of the problems created by the new technologies should be evident from the fact that the world has managed to survive the discovery of nuclear power for more than twenty years. We can examine what has been learned, first to find how scientific and technological knowledge can help solve these problems, and then to discover where science stops and public morality takes over.

MASTERING FALLOUT: THE SCIENTIFIC BASIS

So long as information about fallout was blanketed under military security, the problem of evaluating its effects and what might be done to minimize them was necessarily in the hands of the military, the AEC, and their scientific advisory groups. During that period, the public was assured that there were no hazards from nuclear tests and that no protective action was needed. When secrecy about fallout was lifted, beginning in 1954, the general scientific community became aware of serious inadequacies in the published accounts of the fallout problem and developed a concern about the possible hazards. The next few years were marked by technical (and other) disputes between scientists associated with the government's nuclear test program and the larger group of unaffiliated scientists. Most of these scientific controversies have been resolved, many of them rather rapidly. For example, when Linus Pauling first suggested that carbon-14 from nuclear explosions contributed significantly to the total biological hazard, he was vigorously disputed by government radiation experts. But as soon as a government laboratory made a detailed check of Pauling's calculations, the controversy was settled, for their answers were in substantial agreement with his.

Controversy is nothing new to science; it is common when the available data are insufficient to decide between conflicting points of view. The remedy is more data, and the fallout controversies were useful because they revealed the need for more information about the distribution and effects of fallout radioactivity. When the previously secret AEC reports were made available to scientists, it became apparent that fallout data-gathering was seriously incomplete. Sampling was fairly thorough in the areas near nuclear test sites but very spotty elsewhere in the United States and in the world. Measurements of radiation in foods were scanty. Only a few of the nation's hundreds of milksheds were sampled for strontium-90 at regular intervals; measurements of other foods were scattered and irregular. There were very few measurements of fallout radiation in surface waters. Almost no data were available on certain fallout constituents, especially strontium-89 and iodine-131, the latter omission being seen later as a serious hindrance to understanding the fallout hazard near the Nevada test site.

Many of these inadequacies were pointed out by scientists during the early fallout controversies. It is to the credit of the United States Public Health Service that it responded vigorously to this situation. Beginning in 1957 the USPHS established a program of fallout monitoring which has grown into an elaborate and widespread system of measurements, taken at frequent intervals, of radioactivity in the atmosphere, in surface waters, in the soil, and in foods. The monthly bulletin which the USPHS now issues on these measurements and on data provided by other United States and international agencies is the most detailed information now available about any aspect of environmental contamination.

These new data-gathering systems have strikingly improved our knowledge of fallout, and previous controversies have begun to give way to fact. The initial controversy over the possible hazard from iodine-131 was largely due to the infrequency of the necessary measurements. Because of its short half-life (half of its initial radioactivity decays in only 8 days) iodine-131 can be detected only if measurements are frequent and detailed. When the USPHS monitoring system went into operation it became apparent that each nuclear test in the atmosphere was accompanied by a brief but intense introduction of iodine-131 radioactivity into the food chain. Direct measurements of radio-iodine in milk gave fairly precise calculations of the resultant exposure, especially to children. It then became evident, for the first time, that in the continental United States iodine-131 is responsible for the most intense human exposure to radioactivity from fallout during testing. More important, it became possible to warn milk producers of the hazard and to devise relatively simple countermeasures, such as temporary diversion of milk supplies from the market to bring this hazard under control.

The successful interpretation of such monitoring data was possible only because the source of the contamination was clearly established. Every nuclear test in the atmosphere, by the United States and other nations, is recorded as to size, time, and place. United States laws also require reporting of other possible sources of radioactive contamination, such as nuclear reactors. Such a detailed registry of sources, combined with intensive and widespread monitoring, tells us a great deal about how radioactivity spreads into the environment, and how its hazards can be minimized.

Nevertheless, even with these improvements controversies about fallout persisted. These were centered in the establishment of standards of acceptable exposure. When the fallout issue first arose, the only existing radiation standards were those designed for industrial protection, which were not immediately applicable to situations in which whole continents and entire populations were exposed. For this reason, and because the levels of radioactivity due to fallout were very much lower than those encountered in industry, there was no agreement on what standards ought to apply.

A particularly important lack was the absence of a clearcut theory of biological radiation damage. One theory suggested that repair processes might occur, thus protecting the tissues from any permanent damaging effects from exposure to very low levels of radiation. This approach leads to the concept of a threshold dose which must be exceeded if any biological damage is to ensue. In this case a standard of exposure is fairly easily devised simply by setting it below the threshold dose, which can be determined from experiments with laboratory animals and from observations of accidentally irradiated people. In contrast, another concept of radiation damage—the linear theory—held that there was no threshold and that any increment in radiation exposure would proportionally increase the risk of biological damage. In this case there is no absolute way to establish a standard of tolerable exposure. Any exposure must then be regarded as harmful to some degree.

The scientific community has played a decisive role in resolving this conflict. Largely in response to the fallout problem, geneticists carried out elaborate experiments to study hereditary effects at the low radiation levels which approximate those encountered in fallout. Radiation pathologists also pressed their experiments on tissue damage to lower radiation limits. As a result there is now a rather common agreement that the linear theory of radiation damage is the most reasonable guide to radiation standards. The standards adopted by the responsible United States agency, the Federal Radiation Council, are based on this conclusion.

If any increase in radiation exposure, however slight, is accompanied by a comparable increase in the risk of medically undesirable effects and there is no "safe level" of radiation, how can one determine what dosage is to be tolerated? This judgment requires a balance between the risk associated with a given dosage and some possibly countervailing benefit. The Federal Radiation Council explicitly adopted this position in 1960:

> If . . . beneficial uses were fully exploited without regard to radiation protection, the resulting biological risk might well be considered too great. Reducing the risk to zero would virtually eliminate any radiation use, and result in the loss of all possible benefits. It is therefore necessary to strike some balance between maximum use and zero risk. In establishing radiation protection standards, the balancing of risk and benefit is a decision involving medical, social, economic, political and other factors. Such a balance cannot be made on the basis of a precise mathematical formula but must be a matter of informed judgment.

However, in the actual application of these standards to the fallout problem there has been considerable confusion. When, as a result of atmospheric nuclear tests by the United States and the U.S.S.R. in 1962, the amount of iodine-131 in the milk supplies of several states approached the level which, according to the Federal Radiation Council, required preventive action, local health officials took what they believed to be appropriate measures. In Utah, Wisconsin, and Minnesota, the state departments of health asked farmers to divert fresh milk from the market, so that there would be time for the iodine-131 to decay to acceptable levels before the milk was used for human consumption. But this action was opposed by the United States Secretary of Health, Education and Welfare (who also serves as chairman of the Federal Radiation Council). He stated that the Council's radiation-exposure standards were applicable only to "normal peace time conditions," and according to him these conditions did not include nuclear testing. This interpretation meant that nuclear testing constituted an additional factor not included among the "social, economic, political and other factors" which entered into the Council's original calculation. The Secretary's action can only mean that, in his view, the value of nuclear testing to the nation warranted some increase in the acceptable medical risk from iodine-131. This is a clear illustration of how the seemingly technical questions of environmental pollution very quickly extend beyond the realm of science.

MASTERING PESTICIDES: THE SCIENTIFIC BASIS

Quite similar problems are associated with other forms of environmental pollution, of which the extensive killing of fish near the mouth of the Mississippi River is an example. Beginning in 1957, sugar-cane and cotton farmers in the Mississippi Valley began to spray their crops with a new pesticide, endrin, which is a chemical relative of DDT. Several years later large numbers of dead fish began to appear near the mouth of the river, and set off an intensive investigation and considerable controversy.

Despite deep-seated disagreement between the disputing parties, two facts are clear and acknowledged: Many fish have died and the Mississippi River contains detectable quantities of several insecticides and related organic compounds. The issue is whether the insecticides are the cause of the fish kills and are a hazard to human health, and if so, what should be done about it.

An important argument centers around the possible harm to humans from the very small amounts of pesticide residues present in edible fish and in drinking water taken from the river. One side points to the fact that some laboratory animals exhibit no toxic effects unless exposed to pesticide concentrations many times greater than those due to river pollution. The other side points out that we have inadequate data on the effects of such small concentrations of pesticides on laboratory animals exposed for long periods of time. The possible effects of chronic, long-term exposure of humans to low concentrations of pesticides are also unknown, because, in a sense, the necessary experiment has only just begun.

This issue is the same one that arose in connection with the fallout problem ten years ago, and the same solution is indicated. Estimates of the hazard must be based on the assumption that any increase in exposure results in a proportional risk to the total living population of the biosphere. Like radiation, many of the new synthetic substances act on basic biochemical processes that occur in some form in all living things; therefore some effect on all forms of life must be anticipated. Since some of these pollutants appear to increase the incidence of cancer and the rate of mutation, it is entirely possible that, like radiation, they act on the cell's system of inheritance. Such changes in inheritance may persist in the offspring of the affected organism. The changes are thereby perpetuated and result in an additive risk of eventual biological harm. Moreover, since the biological system exposed to the pesticides is very large and complex, the probability that any increase in contamination will lead to a new point of attack somewhere in this intricate system cannot be ignored. Finally, because the toxic effects of many organic pollutants, like those of radiation, may appear only after a delay of many years, extreme caution ought to be the rule in the early use of pesticides and other novel substances that contaminate the environment.

The very presence in the Mississippi River of substances known to be toxic to fish at low concentrations and to mammals at higher concentrations must be regarded as a definite risk to any biological population exposed to it. The only feasible way to judge the significance of this contamination is to estimate the risks, compare them with the benefits associated with the use of the pesticides, and strike a balance between risk and benefit that will be acceptable to the public. This means that we must know the sources of the contaminants and determine, for example, whether the operation that causes their appearance in the river water is the spraying of corn and cotton crops in the river valley or the activity of riverside plants which manufacture pesticides. This, in turn, will require us to adopt the practice of registering not only the manufacture but also the distribution and use of large amounts of pesticides. Without such a registry there may be no way to determine the source of the contaminants. Finally, against the benefits involved must be balanced the economic losses to the river fisheries and the possible but still unknown hazards to the health of people who absorb the pesticides by eating fish or drinking water taken from the river.

A similar approach is, I believe, equally applicable to most other pollution problems. Since they are all large-scale effects and influence a wide variety of living organisms, on statistical grounds alone it is probable that the smallest detectable pollutant level represents some hazard, however slight, and that the risk will increase roughly with the level. Until known risks can be balanced against specific benefits, no meaningful action is possible. But in the absence of such action the rule of prudence, which is demanded by the unknown long-term hazards, requires that extreme caution be exercised in continued use of these agents.

RISK VERSUS BENEFIT

It appears, then, that the problems of environmental pollution require a common approach: the principle of balancing risk against benefit. The risk can be determined by estimating the number of people exposed to the pollutant, the amounts which they may be expected to absorb, and the physical harm that might result. The benefit can be determined by estimating the economic, political, or social gains expected from the operation which produces the pollutant and the possibilities of substituting less hazardous operations.

Estimations of risk and benefit are proper subjects of scientific and technological analysis. There are scientific means for estimating how many cases of leukemia and of serious congenital defects may result from fallout radiation; such calculations have been reported in great detail by the United Nations Scientific Committee on the Effects of Atomic Radiation. Medical statistics can provide a similar estimate of the amount of respiratory disease that is related to exposure to smog. It should be possible, eventually, to determine what biological risks to humans, birds, or fish are to be expected from a given dosage of DDT.

Determination of the corresponding benefits is more difficult but nevertheless is also within the realm of technological competence. For example, if automobiles are, as they appear to be, a major source of smog in urban areas, it should be possible to evaluate their economic and social importance and to compare it with alternative forms of transport, such as electric trains, which are not smog producers. The economic value of insecticides to the farmer is readily calculable. While no money value can be placed on the benefits to be derived from the development of a new nuclear weapon, it should be possible, it would seem, to determine the necessity of such weapons to the nation, and the importance of nuclear tests to their development.

However difficult the procedures and uncertain the results, all these questions are subject to objective scientific and technological analysis. Presumably scientists who differ in their personal attitudes toward nuclear tests, superhighways, or songbirds could agree, more or less, in their estimates of the relevant benefits and of the associated hazards of fallout, smog, or DDT.

Once the hazards and benefits of new technological innovations become clear, it may be possible to find means to reduce the hazards—at a price. Automobile exhaust emissions can be partially reduced by mechanical devices, which will be required by law on all 1968 models. If we are threatened by accumulating carbon dioxide in the air, engineers can build devices, however expensive, to remove this substance from flue gases. If chemical pesticides are an unwarranted hazard to wildlife or man, we can, after all, stop using them and suffer the sting of the mosquito and the depredations of insects on our crops, while we try to learn enough about environmental biology to develop more natural means of control. If we find strontium-90 intolerable, the nuclear tests that produce fallout can be stopped, and they have in fact been sharply reduced by the test-ban treaty. What is needed is not only the development of technical means for dealing with environmental pollution, but also the willingness to undertake the extra expense and additional inconvenience to prevent the intrusion of pollutants into the environment.

BEYOND THE REALM OF SCIENCE

With the determination of benefits and risk and the development of techniques which improve the balance between them, the applicability of scientific procedure to the problems of environmental contamination comes to an abrupt end. What then remains is a judgment which balances the stated risks against the corresponding benefits. A scientific analysis can perhaps tell us that every nuclear test will probably cause a given number of congenitally deformed births, but no scientific procedures can choose the balance point and tell us how many defective births we ought to tolerate for the sake of a new nuclear weapon.

What is the "importance" of fallout, determined scientifically? Some scientists have stated, with the full dignity of their professional pre-eminence, that the fallout hazard, while not zero, is "trivial." Nevertheless I have seen a minister, upon learning for the first time that acts deliberately performed by his own nation were possibly endangering a few lives in distant lands and a future time, become so incensed at this violation of the biblical injunction against the poisoning of wells as to make an immediate determination to oppose nuclear testing. No science can gauge the relative validity of these conflicting responses to the same facts.

Scientific method cannot determine whether the proponents of urban superhighways or those who complain about the resultant smog are in the right, or whether the benefits of nuclear tests to the national interest outweigh the hazards of fallout. No scientific principle can tell us how to make the choice, which may sometimes be forced upon us by the insecticide problem, between the shade of the elm tree and the song of the robin.

Certainly science can validly describe what is known about the information to be gained from a nuclear experiment, the economic value of a highway, or the hazard of radioactive contamination or of smog. The statement will usually be hedged with uncertainty, and the proper answer may sometimes be "We don't know," but in any case these separate questions do belong within the realm of science. However, the choice of the balance point between benefit and hazard is a value judgment; it is based on ideas of social good, on morality or religion—not on science.

In the "informed judgment" of which the Federal Radiation Council so properly speaks, the scientist can justly claim to be "informed," but he can make no valid claim for a special competence in "judgment." Once the scientific evidence has been stated, or its absence made clear, the establishment of a level of tolerance for a modern pollutant is a *social* problem and must be resolved by social processes. Thus the logic of the scientific problems which are raised by environmental pollution forces the resolution of these issues into the arena of public policy.

If resolutions of the problems created by the recent failures in large-scale technology require social judgments, who is to make these judgments? Obviously scientists must be involved in some way, if only because they have in a sense created the problems. But if these issues require social, political, and moral judgments, then they must also somehow reflect the demands, opinions, and ethics of citizens generally. Because new experiments and technological processes are so costly that the government must often pay for them, and because government officials mediate many social decisions, the government and its administrators are also involved. What are the proper roles of scientist, citizen, and administrator?

THE SCIENTIST'S ROLE: TWO APPROACHES

Since World War II scientists have become deeply concerned with public affairs. We are all acutely aware that our work, our ideas, and our daily activities impinge with a frightening immediacy on national politics, on international conflicts, on the planet's fate as a human habitation.

Scientists have tried to live with these responsibilities in a number of ways. Sometimes, in moments of impending crisis, we are aware only that the main outcome of science is that the planet has become a kind of colossal, lightly triggered time bomb. Then all we can do is to issue an anguished cry of warning. In calmer times we try to grapple with the seemingly endless problems of unraveling the tangle of nuclear physics, seismology, electronics, radiation biology, ecology, sociology, normal and pathological psychology, which, added to the crosscurrents of local, national, and international politics, has become the frightful chaos that goes under the disarming euphemism "public affairs."

Many scientists have studied the technology of the new issues and have mastered their vocabulary: megatonnage, micromicrocuries, threshold dose, and all the rest of the new technical terms. Nuclear physicists have struggled to learn the structure of the chromosome and how cows give milk. Biologists have returned to long-discarded textbooks of freshman physics.

A good deal of the scientist's concern for public issues may be generated by a sense of responsbility for the events which have converted nuclear energy from a laboratory experiment into the force which has almost alone molded the course of human events since 1945. It was a group of scientists who, fearful of the consequences of the possible development of nuclear weapons by the Nazis, conducted a strenuous campaign to convince the American government that they should be achieved in the United States first. As it turned out, Germany never succeeded in achieving an atomic bomb, and the Allies won the war against Germany without using it. Many of the scientists who worked on the United States atomic bomb were relieved to know that the threat which motivated them was gone and that the new force need never be used for destruction. But over their objections the weapon was turned against Japan, an enemy known to lack atomic arms. The human use of nuclear energy began with two explosions which took several hundred thousand lives; from this violent birth it has since grown into a destructive force of suicidal dimensions.

I believe that it is largely the weight of this burden which has caused the scientific community, since the end of the war, to examine with great care the interactions between science and society, to define the scientist's responsibilities to society, and to seek useful ways to discharge them.

For some time there has been a division of opinion within the scientific community on what responsibility the scientist has toward the social uses of his work. Some scientists have been guided by the idea that it is the scientist's duty to pursue knowledge of nature for its own sake without regard to social consequences. They believe that scientists, as scientists, have no special responsibility to foster any particular solution of the social issues that may result from their discoveries. They cling to the objectivity of the laboratory and try to keep their political views separate from their scientific duties. To other scientists such rigorous objectivity seems to imply a disregard for the nation's defense, or for the enormous destructiveness of nuclear war, or for the numerous ways in which science can serve human welfare. They seek to play a part in directing the power that they help to create.

The second of these positions is relatively new and originates in scientists' intense concern with such dangerous issues as nuclear war. The rationale of this position appears to be approximately the following: Scientists have a particular moral responsibility to counter the evil consequences of their works. They are also in possession of the relevant technical facts essential to an understanding of the major public issues which trouble the world. Since scientists are trained to analyze the complex forces at work in such issues, they have an ability for rational thought which renders them to some degree detached from the emotions that encumber the ordinary citizen's views of these calamitous issues. What is more, some assert, the scientist is now in a particularly favorable position to be heard—by government executives, Congressional committees, the press, and the people at large—and therefore has important opportunities to influence social decisions.

This position, it will be noted, is quite neutral politically. It can justify strong statements for or against disarmament, civil defense, nuclear testing, or the space program. Such arguments in support of the scientist's inherent claim to the right of political leadership are the implicit background of a number of practical activities by scientists. These include publication of petitions and newspaper advertisements, political lobbying, and the operation of scientific auxiliaries to the major political parties.

This general approach has become fairly common in the scientific community, in contrast to the viewpoint, more predominant in the past, that a scientist's political role should be exercised apart from his professional one. One result is a growing tendency for considerations of scientific issues to appear with a strong admixture of political views. Witness the following examples from recent discussions, by scientists, of space research:

"On solid scientific grounds, on the basis of popular appeal, and in the interests of our prestige as a peace-loving nation capable of great scientific enterprise, exobiology's goals of finding and exploring extra-territorial life should be acclaimed as the top-priority scientific goal of our space program."

At the 1963 Senate hearings referred to in Chapter 4, a witness who has been a leader in developing the nation's scientific space program said, following an exposition of the scientific values of space research:

"Our goals in space provide to our Nation that spirit and momentum that avoids our collapse into the easy-going days that tolerate social abuse."

In the same vein, I have seen a statement in which a number of distinguished scientists argued for a particular space experiment partly by supporting its scientific value, and partly by describing its special usefulness as a vehicle of international propaganda for the United States.

What is the harm in this approach? Why shouldn't scientists make effective use of their newly won position in society? Why not exploit political interests to further a scientific goal?

One danger has already been discussed—that the capability of science to understand nature, and to guide our efforts to control it, is being damaged by the pressure of political goals. And it is this capability alone that stands between us and man-made catastrophe. Clearly, no matter what else they do, scientists dare not act in such a way as to compromise the integrity of science or to damage its capability to seek the truth. But the notion that scientists have some special aptitude for the judgment of social issues—even of those which are due to the progress of science—runs a grave risk of damaging the integrity of science and public confidence in it. If two protagonists claim to know *as scien-*

tists, through the merits of the methods of science, the one that nuclear testing is essential to the national interest, the other that it is destructive of the national interest, where lies the truth? I know from repetitious experience that the one question about fallout and nuclear war, and about the pesticide controversy, which most troubles the thoughtful citizen is: "How do I know which scientist is telling the truth?"

This is a painfully revealing question. It tells us that the public is no longer certain that scientists—all of them—"tell the truth," for otherwise why the question? The citizen has begun to doubt what he used to take for granted—that science is closely connected to the truth.

Now it seems to me that the citizen's confidence in the objectivity of science cannot be destroyed without disastrous consequences. We cannot really expect citizens, in general, to be capable of performing an independent check on the accuracy and validity of all of a scientist's statements about scientific matters. I am fully convinced that the citizen can and must study and come to understand the underlying facts about modern technological problems. But the citizen cannot check the calculations of the path of a rocket to the moon or question the validity of the law of radioactive decay. These conclusions he can accept, but only if he knows that they *are* subject to the scrutiny of scientists who will finally accept, reject, or modify them. The citizen must take a good deal of science as established by the simple fact that scientists agree about it. It is therefore inevitable that unresolvable disagreement among scientists will erode—and rightly so—the citizen's confidence in the ability of science to get at the truth.

When scientists voice their social judgments with the same authority that attaches to their professional pronouncements, the citizen is bound to confuse the inevitable and insolvable disagreements with scientific disputes. If scientists attach to their scientific conclusions those political views or social judgments which happen to provide support for these conclusions, scientific objectivity inevitably comes under a cloud.

In my opinion the notion that, because the world is dominated by science, scientists have a special competence in public affairs is also profoundly destructive of the democratic process. If we are guided by this view, science will not only create issues but also shield them from the customary processes of administrative decision-making and public judgment.

Nearly every facet of modern life is now so encumbered by a tangled array of nuclear physics, electronics, higher mathematics, and advanced biology as to interpose an apparently insuperable barrier between the citizen, the legislator, the administrator, and the major public issues of the day. No one seems to be wholly exempt from this estrangement. When, during the Congressional hearings on the nuclear-test-ban treaty, Senator Kuchel was confronted with a bewilderingly technical argument, he said in desperation to a scientist member of the AEC, "Let me put my tattered senatorial toga over your shoulders for a moment." When President Kennedy was questioned regarding government policy on the Starfish nuclear explosion, he was forced to fall back on the opinion of a scientist and closed the discussion with the remark, "After all, it's Dr. Van Allen's belt." Confronted by such examples, a citizen is likely to conclude that he must have a Ph.D. to support his judgments about nuclear war or the pesticide problem, or else be governed by the judgment of those who do.

The impact of modern science on public affairs has generated a nearly paralyzing paradox. Despite their origin in scientific knowledge and technological achievements (and failures) the issues created by the advance of science can only be resolved by moral judgment and political choice. But those who in a democratic society have the duty to make these decisions—legislators, government officials, and citizens generally—are often unable to perceive the issues behind the enveloping cloud of science and technology. And if those who have the special knowledge to comprehend the issues—the scientists—arrogate to themselves a major voice in the decision, they are likely to aggravate the very threats to the integrity of science which have helped to generate the problems in the first place.

A NEW APPROACH: AN INFORMED CITIZENRY

There is no single magic word that solves this puzzle. But the same interlocking factors, looked at a little differently, do offer a solution: the scientist does have an urgent duty to *help* society solve the grave problems that have been created by the progress of science. But the problems are social and must be solved by social processes. In these processes the scientist has one vote, and no claim to leadership beyond that given to any person who has the gift of moving his fellow men. But the citizen and the government official whose task it is to make the judgments cannot do so in the absence of the necessary facts and relevant evaluations. Where these are matters of science, the scientist as the custodian of this knowledge has a profound duty to impart as much of it as he can to his fellow citizens. But in doing so, he must guard against false pretensions and avoid claiming for science that which belongs to the conscience. By this means scientists can place the decisions on the grave issues which they have helped to create in the proper hands—the hands of an informed citizenry.

This is the view of the scientist's social responsibilities which was first developed by the AAAS Committee on Science in the Promotion of Human Welfare after a painstaking study of the conflict surrounding science and public policy. In its first report this committee said:

> In sum, we conclude that the scientific community should on its own initiative assume an obligation to call to public attention those issues of public policy which relate to science, and to provide for the general public the facts and estimates of the effects of alternative policies which the citizen must have if he is to participate intelligently in the solution of these problems. A citizenry thus informed is, we believe, the chief assurance that science will be devoted to the promotion of human welfare.

This view is plausible in its logic and laudable in its democratic intent. But does it work? An "informed citizenry" is the long-standing, and rarely attained, goal of social reformers. Most citizens, it can be argued, are unlikely to surmount the formidable difficulties involved in learning about fallout, pesticides, and air pollution. A scientist, once given a platform, might be unable to restrict himself to "the facts and estimates of the effects of alternative policies" and to refrain from also exhorting his audience to whatever policy he has adopted as his own. A scientist who has strong political sensibilities—and many do—may be unable to speak objectively on the data about nuclear war without developing a badly split personality. And by what means can scientists, who have no command over either the public news media or the machinery of government, overcome the governmental self-justification and journalistic inertia which so often impede public knowledge about complex, confused public affairs?

THE COMMITTEE FOR NUCLEAR INFORMATION

Despite all these difficulties a growing number of scientists believe in this approach, because, like the preacher who believed in baptism, they have seen it done. Perhaps the most striking example is the work of the St. Louis Committee for Nuclear Information (CNI), which since 1958 has pioneered in public education about the science-generated issues of public affairs.

Like many other communities in the United States and elsewhere in the world, St. Louis had been troubled and confused by fallout. Nuclear radiation, once a horrifying but (to everyone outside of Hiroshima and Nagasaki) distant prospect had come as close as the morning milk bottle. In 1957 St. Louis became one of six United States cities in which milk supplies were tested for strontium-90 by the Public Health Service. Its citizens learned, at intervals, that there was some number of "micro-microcuries" of strontium-90 in their milk, and that the amount was rising all over the country, with St. Louis often well in the lead.

St. Louis was besieged by numbers. In April 1958 St. Louis milk contained 6.5 micromicrocuries of strontium-90 per liter, but by July the number had increased to 18.7. Health authorities claimed that this amount was "insignificant" because the "maximum permissible concentration" (MPC), a level below which no medical effects were expected, was 80 micromicrocuries per liter. Then testimony at Congressional hearings revealed that, according to the International Commission on Radiological Protection, the MPC for large populations ought to be 30 micromicrocuries per liter and that levels would inevitably rise to that point if nuclear testing continued; in Mandan, North Dakota, milk had already reached the level of 33 micromicrocuries of strontium-90 per liter. Some experts contended that there was no absolutely safe level of radiation and that, according to accepted theories of radiation damage, any increase in radiation caused a proportional increase in the risk of medical damage, which might appear as leukemia or bone cancer.

Along with the confusion came concern. Mothers of young children besieged their pediatricians with questions: Was there really any danger? Should they give their children less milk? What about the strontium-90 content of other foods? Who was right about the MPC? Here was a clear need for an "informed citizenry."

Several of us in the St. Louis scientific community decided to try out the idea that scientists could usefully inform the public about such matters. In April 1958 an organization of several hundred St. Louisans was quickly formed. Citizens helped to define the areas of public concern. They also raised money, manned an office, and did the paper work. Physicians studied the medical problems of fallout. Physicists worked out the amounts and kinds of fallout expected from announced nuclear tests. Chemists concerned themselves with the distribution of the artificial radioisotopes of fallout in the environment. Biologists traced the passage of strontium-90 through the food chain and calculated the expected tissue damage. Seminars were held to exchange information; when the scientists had educated themselves they announced that they were available as public speakers on the facts about fallout.

The response was instantaneous—and overwhelming. When it became known that scientists were willing to try to explain the fallout problem, we were besieged with questions. We had to assure citizens that the white spots on their lawn grass were mold, not fallout; that there was no conceivable way to save the world by extending the half-life of radioactive atoms. And every lecture led to demands for more. Many of us became heavily engaged in the community's lecture circuit: PTAs, Lions, Rotarians, forums, television interviews. We discovered religious denominations and parts of our city that we never before knew existed. We saw at first hand the great need for providing the public with the information that every citizen must have if he is to decide for himself what policies our nation and the world should follow with respect to nuclear testing, civil defense, and disarmament.

At first, most of us tended to regard the task of public education as something apart from our professional life, as a kind of civic responsibility for which our general training prepared us. But, as we began to examine the available evidence for ourselves we found that it did not entirely support certain statements by government officials regarding, for example, the fallout hazard. And an alternative interpretation presented to a lay audience, inevitably called forth the questions: "Why shouldn't the public be guided by the views of the government? Aren't the government scientists better informed than anyone else? Shouldn't we simply listen to what they say?"

CNI scientists found that the basic precept of scientific discourse— that data, interpretations, and conclusions be openly published and criticized—provided an effective answer to this difficulty. The story of the iodine-131 exposures near the Nevada nuclear test site is an example. Despite repeated government assurance of the "safety" of persons living near the test site, CNI scientists were not convinced. They obtained data from AEC and USPHS sources and calculated that the probable radiation dose to children in the area during certain tests in 1953 and later might have been as high as one thousand times the MPC. When this information was offered in testimony before the Joint Congressional Commission on Atomic Energy, the AEC responded with an outright and specific rejection of the CNI conclusions, even though an AEC scientist, H. A. Knapp, had just produced an official report which reached the same conclusion.

This specific critique made it possible for CNI scientists to propose specific replies. For example, in defense of the failure to monitor directly for iodine-131 in the region of the test site, the AEC asserted that in the 1952–57 period there were no techniques available for such analyses. In rebuttal CNI scientists cited scientific publications as early as 1948 which described such techniques, and showed that one had actually been used by the AEC in 1953 to study sheep thyroid glands. In another instance, the AEC asserted that iodine-131 might not fall on pastures which are used by dairy cows, so that such fallout would not reach milk and therefore not harm children. In reply CNI produced a table reporting milk-cow census data for the Utah region, which showed that there were 6,000 dairy cows distributed among several hundred farms in the area. Along with a series of detailed criticisms, CNI sent the AEC Division of Operational Safety a letter asserting that "either you or we are wrong" and asking for evidence of CNI's error or an acknowledgment that the group's analysis of the iodine-131 situation in Utah was correct. Although no reply was forthcoming from the AEC, the net impact of the discussion on other interested parties was noteworthy: The U.S. Public Health Service agreed with the CNI proposal for a medical survey of the children in the Utah area, and such a survey began. . . .

Often the data relevant to science-generated public issues are in fact made fully public by government agencies, but in such technical terms as to be meaningless to the general public. In 1959 the Joint Congressional Committee on Atomic Energy held a series of detailed hearings on the consequence of a nuclear war, assuming a plausible size and distribution of the attack. Witness after witness gave details about the death and destruction of a nuclear holocaust. Although the hearings reported the first comprehensive picture of the potential catastrophe which has so strongly shaped our political lives, most newspaper reports gave only brief mention to a few highlights. The information was there, but the public was not well informed.

CNI arranged with the Joint Committee and with some of the witnesses to obtain copies of the testimony as it was delivered. A team of

CNI scientists studied the testimony and prepared summaries of the physical, ecological, and medical effects of the two nuclear bombs that the Joint Committee's assumed nuclear attack had assigned to St. Louis. The members of the group recognized that scientific data on a subject so remote—and so alarming—as a hypothetical nuclear bombing of a city might be difficult to cast into a readable form but believed that the information was nevertheless of commanding importance. After several unsatisfactory drafts had been prepared they decided that the scientific reports could be translated without loss of detail or accuracy into fictional accounts by supposed survivors of the hypothetical bombing of St. Louis, documented with footnotes relating the facts described to the Congressional hearings. The result was the publication in the CNI monthly magazine *Nuclear Information* (now *Scientist and Citizen*) of a distinctive, moving, yet accurate account of what the evidence presented at the Congressional hearings had to say about the effects of a nuclear attack on a city. In addition to the regular circulation of its publication, CNI distributed, on request, some 45,000 copies of this single issue. It was reprinted in the *Saturday Review*, the *St. Louis Post-Dispatch*, the *Houston Post*, the *San Francisco News-Call Bulletin*, and in a dozen other periodicals in the United States, Great Britain, Canada, Mexico, France, Sweden, and Denmark.

Here the CNI scientists had served as translators—bridging the morass of technical terms and scientific abstractions which separates most citizens from the awful fact of nuclear war. And apparently even trained government officials needed CNI's translation service to appreciate what went on at the Congressional hearings: On request CNI distributed 50 copies of the article to the Red Cross, 800 to various branches of the military services, 1,000 to the Food and Drug Administration, and 2,000 to the Department of Health, Education, and Welfare. About 2,500 copies were ordered by state and federal civil defense organizations. The War College and the Air University of the United States Air Force received permission to reprint the bulletin for the use of their graduating classes.

Does such information do any good? Even if informed, what can the citizen do about established government policy? A good test of this question is the story of Project Chariot, a proposal put forward by the AEC in 1958 to blast out a new harbor on the northwest shore of Alaska with hydrogen bombs. Soon after the proposal was announced, a controversy arose regarding radiation effects on local flora and fauna and on the life of the nearby Eskimo villages. CNI asked the AEC for information. The AEC replied that while no reports on the biological hazards were available, a number of biologists, whose names were provided, were engaged in such studies under AEC sponsorship. CNI wrote to these biologists for information. Several at the University of Alaska replied with copies of reports that they had earlier submitted to the AEC. Study of these soon showed that the Alaskan investigations had uncovered a series of important biological difficulties in Project Chariot.

For example, the AEC had apparently selected the site of the proposed explosion because it was in the center of a long reach of uninhabited shoreline. But what had not been taken into account was that inhabitants from two Eskimo villages, although distant from the site, intensively hunted and gathered food in the "empty" region; vast food-gathering areas are essential for survival in arctic areas.

CNI invited the Alaskan biologists to prepare articles for publication in *Nuclear Information* [*Scientist and Citizen*]. To these articles were added studies by CNI physicists of some of the problems involved in such earth-moving projects. The result was an extensive analysis of Project Chariot, which showed, among other things, that it would seriously endanger the food supply of the local Eskimo settlements. At about the same time the AEC issued a report on the state of the project, which gave scant attention to these biological difficulties. For several years Alaskans had been assured by AEC representatives that there was no danger from the explosions. Now it appeared that there was, on scientific grounds, another side to the story.

But could such information really be brought to bear on a governmental apparatus that had already committed millions of dollars to the project? The people most concerned were the inhabitants of the several isolated, very primitive Eskimo settlements along the west coast of Alaska. Few of the villagers could read or speak English. They even lacked a written language of their own. But with the nuclear age we also have the tape recorder, with consequences that have been reported by Dr. W. H. Pruitt, one of the Alaskan biologists who studied the ecological effects of the Chariot proposal:

The tape recorder is one of the greatest inventions as far as the Eskimo culture is concerned. Every village has several tape recorders. . . . There is a constant traffic in tapes from one village to another. . . . The Chariot information and the material contained in the Alaska Conservation Society and CNI bulletins got into the tapes and it literally swept the Arctic coast, from Kakhtovik all the way down to Nome and below. . . . I recall, also, meeting an Eskimo driving a dog team on the trail one time, and, by golly, he had a copy of the CNI bulletin tucked inside his parka.

It is risky to claim a particular cause for any given event in political life, but the record does show that the Chariot proposal met little opposition until independent scientists of the Alaska Conservation Society and CNI began to explain some of the biological consequences —and that the project was finally killed.

Public information developed by independent scientists has figured in a number of controversies over nuclear projects. When the State of Massachusetts proposed to establish a nuclear-waste-processing plant on Cape Cod, scientists at the Marine Biological Laboratory at Woods Hole assembled information showing that special hazards would result from the geological situation peculiar to the region. After public hearings at which these facts were aired, the project was abandoned. The proposal to build a large nuclear reactor at Bodega Bay in California ran into stiff opposition based largely on evidence that the reactor would be so close to the San Andreas earthquake fault as to raise a serious hazard of an earthquake-induced radiation catastrophe. Many people were surprised at the ease with which the 1963 test-ban treaty was approved by the United States Senate. Several observers have noted a possible connection with the numerous letters received by Senators from housewives and mothers who not only wanted the treaty approved but could cite serious scientific grounds for their belief that it would help reduce the medical hazards from fallout.

In a surprising way CNI's program has turned out to be a two-way street. While citizens are usually on the receiving end of information provided by scientists, in at least one important instance in connection with fallout citizens have themselves produced new scientific information. In 1958 the prominent American biochemist Herman Kalckar published a paper in the international scientific journal *Nature* on the problem of assessing the strontium-90 uptake in children's bones. The importance of such information was widely recognized, since there was no basis from previous experience for an accurate prediction of how much of the strontium-90 in the diet would remain in the bones—where it might, if sufficiently concentrated, in time induce leukemia or bone tumors. Methods were available for strontium-90 analysis of bones, but since the analysis could only be made on corpses the data were scattered and incomplete. Dr. Kalckar pointed out that strontium-90 is also deposited in the teeth, that the milk teeth are shed after a time and could easily be analyzed for strontium-90, and that the results could be directly correlated with the situation in bone.

The idea was sound but difficult to carry out. A single tooth contains so little strontium-90 that considerable numbers of teeth would be needed to produce statistically significant results. The teeth would need to be collected regularly over a long period in order to make up for the delay of about 5 to 7 years between the time a milk tooth is formed and the time it is shed. When CNI scientists read Dr. Kalckar's paper, they estimated that the project would require the collection of about 50,000 teeth a year for at least ten years to produce useful results.

But CNI members were struck with the idea and enthusiastic about "doing something" about fallout, if only providing scientific information about it. In December 1958 CNI organized the Baby Tooth Survey and with the help of numerous willing mothers and children began collecting teeth. When the project began no one was sure how so many teeth could be collected, nor did CNI have the resources to carry out the elaborate and expensive analyses for strontium-90. Within a year the donation of baby teeth had become a way of life among St. Louis children. By 1966 the Baby Tooth Survey had collected over 200,000 teeth. With CNI's cooperation the School of Dentistry at Washington University established an analytical laboratory, supported by a grant from the U.S. Public Health Services. The laboratory has published the first, and thus far the only, detailed studies of the absorption of strontium-90 by a large population of children through the major period of nuclear testing. It has provided scientists with unique data on this new problem. The children who gave up the traditional visit of the tooth fairy to contribute their baby teeth to science were in this case helping to develop new scientific information for their own future welfare.

Much has been written about the alienation of citizens from the complexities of modern public affairs. Experts have everywhere intruded between the issues and the public. The Jeffersonian concept of an educated, informed electorate appears to be a naïve and distant ideal. But at least in one area—science and technology, on which so much of our future depends—an effort is being made to make the ideal a reality.

Margaret Mead, the anthropologist, has called scientists' efforts to alert citizens to these issues and provide them with information needed for evaluating the benefits and hazards of modern technology "a new social invention." This may turn out to be the one invention of our technological age which can conserve the environment and preserve life on the earth.

ECO-CATASTROPHE! *

PAUL EHRLICH

Dr. Paul R. Ehrlich is an outstanding and outspoken ecologist, and a Professor of Biology at Stanford University. His specialty is population biology, and he is a founder of Zero Population Growth. His hard-hitting realistic book, "The Population Bomb," pulls no punches. In the following essay, entitled *Eco-Catastrophe!*, Professor Ehrlich predicts what our world will be like in ten years if the present course of environmental destruction is allowed to continue.

The end of the ocean came late in the summer of 1979, and it came even more rapidly than the biologists had expected. There had been signs for more than a decade, commencing with the discovery in 1968 that DDT slows down photosynthesis in marine plant life. It was announced in a short paper in the technical journal, Science, but to ecologists it smacked of doomsday. They knew that all life in the sea depends on photosynthesis, the chemical process by which green plants bind the sun's energy and make it available to living things. And they knew that DDT and similar chlorinated hydrocarbons had polluted the entire surface of the earth, including the sea.

But that was only the first of many signs. There had been the final gasp of the whaling industry in 1973, and the end of the Peruvian anchovy fishery in 1975. Indeed, a score of other fisheries had disappeared quietly from over-exploitation and various eco-catastrophes by 1977. The term "eco-catastrophe" was coined by a California ecologist in 1969 to describe the most spectacular of man's attacks on the systems which sustain his life. He drew his inspiration from the Santa Barbara offshore oil disaster of that year, and from the news which spread among naturalists that virtually all of the Golden State's seashore bird life was doomed because of chlorinated hydrocarbon interference with its reproduction. Eco-catastrophes in the sea became increasingly common in the early 1970's. Mysterious "blooms" of previously rare microorganisms began to appear in offshore waters. Red tides—killer outbreaks of a minute single-celled plant—returned to the Florida Gulf coast and were sometimes accompanied by tides of other exotic hues.

It was clear by 1975 that the entire ecology of the ocean was changing. A few types of phytoplankton were becoming resistant to chlorinated hydrocarbons and were gaining the upper hand. Changes in the phytoplankton community led inevitably to changes in the community of zooplankton, the tiny animals which eat the phytoplankton. These changes were passed on up the chains of life in the ocean to the herring, plaice, cod and tuna. As the diversity of life in the ocean diminished, its stability also decreased.

Other changes had taken place by 1975. Most ocean fishes that returned to fresh water to breed, like the salmon, had become extinct, their breeding streams so dammed up and polluted that their powerful homing instinct only resulted in suicide. Many fishes and shellfishes that bred in restricted areas along the coasts followed them as onshore pollution escalated.

By 1977 the annual yield of fish from the sea was down to 30 million metric tons, less than one-half the per capita catch of a decade earlier. This helped malnutrition to escalate sharply in a world where an estimated 50 million people per year were already dying of starvation. The United Nations attempted to get all chlorinated hydrocarbon insecticides banned on a worldwide basis, but the move was defeated by the United States. This opposition was generated primarily by the American petrochemical industry, operating hand in glove with its subsidiary, the United States Department of Agriculture. Together they persuaded the government to oppose the U.N. move—which was not difficult since most Americans believed that Russia and China were more in need of fish products than was the United States. The United Nations also attempted to get fishing nations to adopt strict and enforced catch limits to preserve dwindling stocks. This move was blocked by Russia, who, with the most modern electronic equipment, was in the best position to glean what was left in the sea. It was, curiously, on the very day in 1977 when the Soviet Union announced its refusal that another ominous article appeared in Science. It announced that incident solar radiation had been so reduced by worldwide air pollution that serious effects on the world's vegetation could be expected.

Apparently it was a combination of ecosystem destabilization, sunlight reduction, and a rapid escalation in chlorinated hydrocarbon pollution from massive Thanodrin applications which triggered the ultimate catastrophe. Seventeen huge Soviet-financed Thanodrin plants were operating in underdeveloped countries by 1978. They had been part of a massive Russian "aid offensive" designed to fill the gap caused by the collapse of America's ballyhooed "Green Revolution."

It became apparent in the early '70s that the "Green Revolution" was more talk than substance. Distribution of high yield "miracle" grain seeds had caused temporary local spurts in agricultural production. Simultaneously, excellent weather had produced record harvests. The combination permitted bureaucrats, especially in the United States Department of Agriculture and the Agency for International Development (AID), to reverse their previous pessimism and indulge in an outburst of optimistic propaganda about staving off famine. They raved about the approaching transformation of agriculture in the underdeveloped countries (UDCs). The reason for the propaganda reversal was never made clear. Most historians agree that a combination of utter ignorance of ecology, a desire to justify past errors, and pressure from agro-industry (which was eager to sell pesticides, fertilizers, and farm machinery to the UDCs and agencies helping the UDCs) was behind the campaign. Whatever the motivation, the results were clear. Many concerned people, lacking the expertise to see through the Green Revolution drivel, relaxed. The population-food crisis was "solved."

But reality was not long in showing itself. Local famine persisted in northern India even after good weather brought an end to the ghastly Bihar famine of the mid-'60s. East Pakistan was next, followed by a resurgence of general famine in northern India. Other foci of famine rapidly developed in Indonesia, the Philippines, Malawi, the Congo, Egypt, Colombia, Ecuador, Honduras, the Dominican Republic, and Mexico.

Everywhere hard realities destroyed the illusion of the Green Revolution. Yields dropped as the progressive farmers who had first accepted the new seeds found that their higher yields brought lower prices—effective demand (hunger plus cash) was not sufficient in poor countries to keep prices up. Less progressive farmers, observing this, refused to make the extra effort required to cultivate the "miracle" grains. Transport systems proved inadequate to bring the necessary fertilizer to the fields where the new and extremely fertilizer-sensitive grains were being grown. The same systems were also inadequate to move produce to markets. Fertilizer plants were not built fast enough, and most of the underdeveloped countries could not scrape together funds to purchase supplies, even on concessional terms. Finally, the inevitable happened, and pests began to reduce yields in even the most carefully cultivated fields. Among the first were the famous "miracle rats" which invaded Philippine "miracle rice" fields early in 1969. They were quickly followed by many insects and viruses, thriving on the relatively pest-susceptible new grains, encouraged by the vast and dense plantings, and rapidly acquiring resistance to the chemicals used against them. As chaos spread until even the most obtuse agriculturists and economists realized that the Green Revolution had turned brown, the Russians stepped in.

In retrospect it seems incredible that the Russians, with the American mistakes known to them, could launch an even more incompetent program of aid to the underdeveloped world. Indeed, in the early 1970's there were cynics in the United States who claimed that outdoing the stupidity of American foreign aid would be physically impossible. Those critics were, however, obviously unaware that the Russians had been busily destroying their own environment for many years. The virtual disappearance of sturgeon from Russian rivers caused a great shortage of caviar by 1970. A standard joke among Russian scientists at that time was that they had created an artificial caviar which was indistinguishable from the real thing—except by taste. At any rate the Soviet Union, observing with interest the progressive deterioration of relations between the UDCs and the United States, came up with a solution. It had recently developed what it claimed was the ideal insecticide, a highly lethal chlorinated hydrocarbon complexed with a special agent for penetrating the external skeletal armor of insects. Announcing that the new pesticide, called Thanodrin, would truly produce a Green Revolution, the Soviets entered into negotiations with various UDCs for the construction of massive Thanodrin factories. The USSR would bear all the costs; all it wanted in return were certain trade and military concessions.

It is interesting now, with the perspective of years, to examine in some detail the reasons why the UDCs welcomed the Thanodrin plan with such open arms. Government officials in these countries ignored the protests of their own scientists that Thanodrin would not solve the problems which plagued them. The governments now knew that the basic cause of their problems was overpopulation, and that these problems had been exacerbated by the dullness, daydreaming, and cupidity endemic to all governments. They knew that only population control and limited development aimed primarily at agriculture could have spared them the horrors they now faced. They knew it, but they were not about to admit it. How much easier it was simply to accuse the Americans of failing to give them proper aid; how much simpler to accept the Russian panacea.

And then there was the general worsening of relations between the United States and the UDCs. Many things had contributed to this. The situation in America in the first half of the 1970's deserves our close scrutiny. Being more dependent on imports for raw materials than the Soviet Union, the United States had, in the early 1970's, adopted more and more heavy-handed policies in order to insure continuing supplies. Military adventures in Asia and Latin America had further lessened the international credibility of the United States as a great defender of freedom—an image which had begun to deteriorate rapidly during the pointless and fruitless Viet-Nam conflict. At home, acceptance of the carefully manufactured image lessened dramatically, as even the more romantic and chauvinistic citizens began to understand the role of the military and the industrial system in what John Kenneth Galbraith had aptly named "The New Industrial State."

At home in the USA the early '70s were traumatic times. Racial violence grew and the habitability of the cities diminished, as nothing substantial was done to ameliorate either racial inequities or urban blight. Welfare rolls grew as automation and general technological progress forced more and more people into the category of "unemployable." Simultaneously a taxpayers' revolt occurred. Although there was not enough money to build the schools, roads, water systems, sewage systems, jails, hospitals, urban transit lines, and all the other amenities needed to support a burgeoning population, Americans refused to tax themselves more heavily. Starting in Youngstown, Ohio in 1969 and followed closely by Richmond, California, community after community was forced to close its schools or curtail educational operations for lack of funds. Water supplies, already marginal in quality and quantity in many places by 1970, deteriorated quickly. Water rationing occurred in 1723 municipalities in the summer of 1974, and hepatitis and epidemic dysentery rates climbed about 500 per cent between 1970–1974.

Air pollution continued to be the most obvious manifestation of environmental deterioration. It was, by 1972, quite literally in the eyes of all Americans. The year 1973 saw not only the New York and Los Angeles smog disasters, but also the publication of the Surgeon General's massive report on air pollution and health. The public had been partially prepared for the worst by the publicity given to the U.N. pollution conference held in 1972. Deaths in the late '60s caused by smog were well known to scientists, but the public had ignored them because they mostly involved the early demise of the old and sick rather than people dropping dead on the freeways. But suddenly our citizens were faced with nearly 200,000 corpses and massive documentation that they could be the next to die from respiratory disease. They were not ready for that scale of disaster. After all, the U.N. conference had not predicted that accumulated air pollution would make the planet uninhabitable until almost 1990. The population was terrorized as TV screens became filled with scenes of horror from the disaster areas. Especially vivid was NBC's coverage of hundreds of unattended people choking out their lives outside of New York's hospitals. Terms like nitrogen oxide, acute bronchitis and cardiac arrest began to have real meaning for most Americans.

The ultimate horror was the announcement that chlorinated hydrocarbons were now a major constituent of air pollution in all American cities. Autopsies of smog disaster victims revealed an average chlorinated hydrocarbon load in fatty tissue equivalent to 26 parts per million of DDT. In October, 1973, the Department of Health, Education and Welfare announced studies which showed unequivocally that increasing death rates from hypertension, cirrhosis of the liver, liver cancer and a series of other diseases had resulted from the chlorinated hydrocarbon load. They estimated that Americans born since 1946 (when DDT usage began) now had a life expectancy of only 49 years, and predicted that if current patterns continued, this expectancy would reach 42 years by 1980, when it might level out. Plunging insurance stocks triggered a stock market panic. The president of Velsicol, Inc., a major pesticide producer, went on television to "publicly eat a teaspoonful of DDT" (it was really powdered milk) and announce that HEW had been infiltrated by Communists. Other giants of the petro-chemical industry, attempting to dispute the indisputable evidence, launched a massive pressure campaign on Congress to force HEW to "get out of agriculture's business." They were aided by the agro-chemical journals, which had decades of experience in misleading the public about the benefits and dangers of pesticides. But by now the public realized that it had been duped. The Nobel Prize for medicine and physiology was given to Drs. J. L. Radomski and W. B. Deichmann, who in the late 1960's had pioneered in the documentation of the long-term lethal effects of chlorinated hydrocarbons. A Presidential Commission with unimpeachable credentials directly accused the agro-chemical complex of "condemning many millions of Americans to an early death." The year 1973 was the year in which Americans finally came to understand the direct threat to their existence posed by environmental deterioration.

And 1973 was also the year in which most people finally comprehended the indirect threat. Even the president of Union Oil Company and several other industrialists publicly stated their concern over the reduction of bird populations which had resulted from pollution by DDT and other chlorinated hydrocarbons. Insect populations boomed because they were resistant to most pesticides and had been freed, by the incompetent use of those pesticides, from most of their natural enemies. Rodents swarmed over crops, multiplying rapidly in the absence of predatory birds. The effect of pests on the wheat crop was especially disastrous in the summer of 1973, since that was also the year of the great drought. Most of us can remember the shock which greeted the announcement by atmospheric physicists that the shift of the jet stream which had caused the drought was probably permanent. It signalled the birth of the Midwestern desert. Man's air-polluting activities had by then caused gross changes in climatic patterns. The news, of course, played hell with commodity and stock markets. Food prices skyrocketed, as savings were poured into hoarded canned goods. Official assurances that food supplies would remain ample fell on deaf ears, and even the government showed signs of nervousness when California migrant field workers went out on strike again in protest against the continued use of pesticides by growers. The strike burgeoned into farm burning and riots. The workers, calling themselves "The Walking Dead," demanded immediate compensation for their shortened lives, and crash research programs to attempt to lengthen them.

It was in the same speech in which President Edward Kennedy, after much delay, finally declared a national emergency and called out the National Guard to harvest California's crops, that the first mention of population control was made. Kennedy pointed out that the United States would no longer be able to offer any food aid to other nations and was likely to suffer food shortages herself. He suggested that, in view of the manifest failure of the Green Revolution, the only hope of the UDCs lay in population control. His statement, you will recall, created an uproar in the underdeveloped countries. Newspaper editorials accused the United States of wishing to prevent small countries from becoming large nations and thus threatening American hegemony. Politicians asserted that President Kennedy was a "creature of the giant drug combine" that wished to shove its pills down every woman's throat.

Among Americans, religious opposition to population control was very slight. Industry in general also backed the idea. Increasing poverty in the UDCs was both destroying markets and threatening supplies of raw materials. The seriousness of the raw material situation had been brought home during the Congressional Hard Resources hearings in 1971. The exposure of the ignorance of the cornucopian economists had been quite a spectacle—a spectacle brought into virtually every American's home in living color. Few would forget the distinguished geologist from the University of California who suggested that economists be legally required to learn at least the most elementary facts of geology. Fewer still would forget that an equally distinguished Harvard economist added that they might be required to learn some economics, too. The overall message was clear: America's resource situation was bad and bound to get worse. The hearings had led to a bill requiring the Departments of State, Interior, and Commerce to set up a joint resource procurement council with the express purpose of "insuring that proper consideration of American resource needs be an integral part of American foreign policy."

Suddenly the United States discovered that it had a national consensus: population control was the only possible salvation of the underdeveloped world. But that same consensus led to heated debate. How could the UDCs be persuaded to limit their populations, and should not the United States lead the way by limiting its own? Members of the intellectual community wanted America to set an example. They pointed out that the United States was in the midst of a new baby boom: her birth rate, well over 20 per thousand per year, and her growth rate of over one per cent per annum were among the very highest of the developed countries. They detailed the deterioration of the American physical and psychic environments, the growing health threats, the impending food shortages, and the insufficiency of funds for desperately needed public works. They contended that the nation was clearly unable or unwilling to properly care for the people it already had. What possible reason could there be, they queried, for adding any more? Besides, who would listen to requests by the United States for population control when that nation did not control her own profligate reproduction?

Those who opposed population controls for the U.S. were equally vociferous. The military-industrial complex, with its all-too-human mixture of ignorance and avarice, still saw strength and prosperity in numbers. Baby food magnates, already worried by the growing nitrate pollution of their products, saw their market disappearing. Steel manufacturers saw a decrease in aggregate demand and slippage for that holy of holies, the Gross National Product. And military men saw, in the growing population-food-environment crisis, a serious threat to their carefully nurtured Cold War. In the end, of course, economic arguments held sway, and the "inalienable right of every American couple to determine the size of its family," a freedom invented for the occasion in the early '70s, was not compromised.

The population control bill, which was passed by Congress early in 1974, was quite a document, nevertheless. On the domestic front, it authorized an increase from 100 to 150 million dollars in funds for "family planning" activities. This was made possible by a general feeling in the country that the growing army on welfare needed family planning. But the gist of the bill was a series of measures designed to impress the need for population control on the UDCs. All American aid to countries with overpopulation problems was required by law to consist in part of population control assistance. In order to receive any assistance each nation was required not only to accept the population control aid, but also to match it according to a complex formula. "Overpopulation" itself was defined by a formula based on U.N. statistics, and the UDCs were required not only to accept aid, but also to show progress in reducing birth rates. Every five years the status of the aid program for each nation was to be re-evaluated.

The reaction to the announcement of this program dwarfed the response to President Kennedy's speech. A coalition of UDCs attempted to get the U.N. General Assembly to condemn the United States as a "genetic aggressor." Most damaging of all to the American cause was the famous "25 Indians and a dog" speech by Mr. Shankarnarayan, Indian Ambassador to the U.N. Shankarnarayan pointed out that for several decades the United States, with less than six per cent of the people of the world had consumed roughly 50 per cent of the raw materials used every year. He described vividly America's contribution to worldwide environmental deterioration, and he scathingly denounced the miserly record of United States foreign aid as "unworthy of a fourth-rate power, let alone the most powerful nation on earth."

It was the climax of his speech, however, which most historians claim once and for all destroyed the image of the United States. Shankarnarayan informed the assembly that the average American family dog was fed more animal protein per week than the average Indian got in a month. "How do you justify taking fish from protein-starved Peruvians and feeding them to your animals?" he asked. "I contend," he concluded, "that the birth of an American baby is a greater disaster for the world than that of 25 Indian babies." When the applause had died away, Mr. Sorensen, the American representative, made a speech which said essentially that "other countries look after their own self-interest, too." When the vote came, the United States was condemned.

This condemnation set the tone of U.S.-UDC relations at the time the Russian Thanodrin proposal was made. The proposal seemed to offer the masses in the UDCs an opportunity to save themselves and humiliate the United States at the same time; and in human affairs, as we all know, biological realities could never interfere with such an opportunity. The scientists were silenced, the politicians said yes, the Thanodrin plants were built, and the results were what any beginning ecology student could have predicted. At first Thanodrin seemed to offer excellent control of many pests. True, there was a rash of human fatalities from improper use of the lethal chemical, but, as Russian technical advisors were prone to note, these were more than compensated for by increased yields. Thanodrin use skyrocketed throughout the underdeveloped world. The Mikoyan design group developed a dependable, cheap agricultural aircraft which the Soviets donated to the effort in large numbers. MIG sprayers became even more common in UDCs than MIG interceptors.

Then the troubles began. Insect strains with cuticles resistant to Thanodrin penetration began to appear. And as streams, rivers, fish culture ponds and onshore waters became rich in Thanodrin, more fisheries began to disappear. Bird populations were decimated. The sequence of events was standard for broadcast use of a synthetic pesticide: great success at first, followed by removal of natural enemies and development of resistance by the pest. Populations of crop-eating insects in areas treated with Thanodrin made steady comebacks and soon became more abundant than ever. Yields plunged, while farmers in their desperation increased the Thanodrin dose and shortened the time between treatments. Death from Thanodrin poisoning became common. The first violent incident occurred in the Canete Valley of Peru, where farmers had suffered a similar chlorinated hydrocarbon disaster in the mid-'50s. A Russian advisor serving as an agricultural pilot was assaulted and killed by a mob of enraged farmers in January, 1978. Trouble spread rapidly during 1978, especially after the word got out that two years earlier Russia herself had banned the use of Thanodrin at home because of its serious effects on ecological systems. Suddenly Russia, and not the United States, was the *bête noir* in the UDCs. "Thanodrin parties" became epidemic, with farmers, in their ignorance, dumping carloads of Thanodrin concentrate into the sea. Russian advisors fled, and four of the Thanodrin plants were leveled to the ground. Destruction of the plants in Rio and Calcutta led to hundreds of thousands of gallons of Thanodrin concentrate being dumped directly into the sea.

Mr. Shankarnarayan again rose to address the U.N., but this time it was Mr. Potemkin, representative of the Soviet Union, who was on the hot seat. Mr. Potemkin heard his nation described as the greatest mass killer of all time as Shankarnarayan predicted at least 30 million deaths from crop failures due to overdependence on Thanodrin. Russia was accused of "chemical aggression," and the General Assembly, after a weak reply by Potemkin, passed a vote of censure.

It was in January, 1979, that huge blooms of a previously unknown variety of diatom were reported off the coast of Peru. The blooms were accompanied by a massive die-off of sea life and of the pathetic remainder of the birds which had once feasted on the anchovies of the area. Almost immediately another huge bloom was reported in the Indian ocean, centering around the Seychelles, and then a third in the South Atlantic off the African coast. Both of these were accompanied by spectacular die-offs of marine animals. Even more ominous were growing reports of fish and bird kills at oceanic points where there were no spectacular blooms. Biologists were soon able to explain the phenomena: the diatom had evolved an enzyme which broke down Thanodrin; that enzyme also produced a breakdown product which interfered with the transmission of nerve impulses, and was therefore lethal to animals. Unfortunately, the biologists could suggest no way of repressing the poisonous diatom bloom in time. By September, 1979, all important animal life in the sea was extinct. Large areas of coastline had to be evacuated, as windrows of dead fish created a monumental stench.

But stench was the least of man's problems. Japan and China were faced with almost instant starvation from a total loss of the seafood on which they were so dependent. Both blamed Russia for their situation and demanded immediate mass shipments of food. Russia had none to send. On October 13, Chinese armies attacked Russia on a broad front. . . .

A pretty grim scenario. Unfortunately, we're a long way into it already. Everything mentioned as happening before 1970 has actually occurred; much of the rest is based on projections of trends already appearing. Evidence that pesticides have long-term lethal effects on human beings has started to accumulate, and recently Robert Finch, Secretary of the Department of Health, Education and Welfare expressed his extreme apprehension about the pesticide situation. Simultaneously the petrochemical industry continues its unconscionable poison-peddling. For instance, Shell Chemical has been carrying on a high-pressure campaign to seel the insecticide Azodrin to farmers as a killer of cotton pests. They continue their program even though they know that Azodrin is not only ineffective, but often *increases* the pest density. They've covered themselves nicely in an advertisement which states, "Even if an overpowering migration [sic] develops, the flexibility of Azodrin lets you regain control fast. Just increase the dosage according to label recommendations." It's a great game—get people to apply the poison and kill the natural enemies of the pests. Then blame the increased pests on "migration" and sell even more pesticide!

Right now fisheries are being wiped out by over-exploitation, made easy by modern electronic equipment. The companies producing the equipment know this. They even boast in advertising that only their equipment will keep fishermen in business until the final kill. Profits must obviously be maximixed in the short run. Indeed, Western society is in the process of completing the rape and murder of the planet for economic gain. And, sadly, most of the rest of the world is eager for the opportunity to emulate our behavior. But the underdeveloped peoples will be denied that opportunity—the days of plunder are drawing inexorably to a close.

Most of the people who are going to die in the greatest cataclysm in the history of man have already been born. More than three and a half billion people already populate our moribund globe, and about half of them are hungry. Some 10 to 20 million will starve to death *this year*. In spite of this, the population of the earth will increase by 70 million souls in 1969. For mankind has artificially lowered the death rate of the human population, while in general birth rates have remained high. With the input side of the population system in high gear and the output side slowed down, our fragile planet has filled with people at an incredible rate. It took several million years for the population to reach a total of two billion people in 1930, while a *second two billion will have been added by 1975!* By that time some experts feel that food shortages will have escalated the present level of world hunger and starvation into famines of unbelievable proportions. Other experts, more optimistic, think the ultimate food-population collision will not occur until the decade of the 1980's. Of course more massive famine may be avoided if other events cause a prior rise in the human death rate.

Both worldwide plague and thermonuclear war are made more probable as population growth continues. These, along with famine, make up the trio of potential "death rate solutions" to the population problem—solutions in which the birth rate-death rate imbalance is redressed by a rise in the death rate rather than by a lowering of the birth rate. Make no mistake about it, *the imbalance will be redressed.* The shape of the population growth curve is one familiar to the biologist. It is the outbreak part of an outbreak-crash sequence. A population grows rapidly in the presence of abundant resources, finally runs out of food or some other necessity, and crashes to a low level or extinction. Man is not only running out of food, he is also destroying the life support systems of the Spaceship Earth. The situation was recently summarized very succinctly: "It is the top of the ninth inning. Man, always a threat at the plate, has been hitting Nature hard. It is important to remember, however, that NATURE BATS LAST."

THE FALLACY OF
SLUM CLEARANCE
AND THE REMEDY *

WILLIAM AND PAUL PADDOCK

William and Paul Paddock are no ivory tower surveyors but hard-headed realists. Both are eminently qualified to speak with authority in the areas of hunger, overpopulation, and conservation, having spent four decades of work in the hungry nations of the world, one brother as an agronomist and educator and the other as a political officer in the U.S. State Department.

In their book "Hungry Nations," they dispel vague thoughts of miracles to be worked with American money and know-how.

Their most famous book is "Famine: 1975." It is sobering, realistic, frightening, and beautifully documented.

By "fallacy" I mean something embedded in a development program that brings to naught the efforts expended.

The human element—that is, human failing—is not involved here, although the fallacy can lead to a way of thinking or an atmosphere of public opinion that in itself affects economic development goals adversely.

No, the fallacy is endemic within the program. Even before the first step is taken, the first shovel turned, the program can end only in failure. Rather, the specific project can be a success, but the goal of raising a nation's standard of living is foredoomed.

Slum clearance, as included in a nation's development program, is an easy fallacy with which to start. Everyone wants to get the slums of the cities of the hungry countries cleared away. Everyone believes that with the slums gone a lot of social and economic problems will be resolved. All will be happy. All will be well.

Alas, whatever slum clearance may do for places like New York, St. Louis, Manchester and Milan, there is a sad fallacy within this wishful thinking insofar as the hungry nations are concerned.

The low-rent housing and apartment centers presently so favored by most government leaders and American aid officials are not always intended as "slum clearance" in the strict sense of the term. Nevertheless, the aim of these new clusters of buildings is to lessen urban congestion—to decompress the bodies in the city alleys. Thus, it seems proper to include in the overall term of "slum clearance" these new urban developments, both those financed by local governments and those financed by foreign aid allotments.

Nearly every writer on foreign subjects, certainly every prosperous American tourist, at some time weeps over the horrendous slums he finds in his travels.

This same person, most likely, has never in his entire life walked through a true American slum, such as the backwash of Brooklyn, Harlem, Chicago and, really, almost any of our cities. Yet the American slum would be paradise indeed for the unfortunates he sees in the *bidonvilles* of Casablanca, Bogotá, Cairo, Bombay.

Thus, the American writer or tourist who is ignorant of his own slums is doubly shocked when confronted with the squalor abroad. There is no need to repeat here their accounts of the filth, the stinks, the emaciated body propped exhausted against a wall and, worst of all, the despair and hopelessness of the humans abandoned to these hells by the disinterest of their fellow beings, their church and their government.

City slums are not, of course, a modern phenomenon. They have formed a counterpoint in the life of every city in the past. Ancient Rome, famous for so many beautiful palaces, was also famous for its awesome slums—and for the street mobs that erupted from them.

The new element today is that man, for the first time, dreams of cities where slums do not exist. In this "age of rising expectations" slum mobs riot when their hopes are frustrated and, like mindless pixies, stone the American embassy but not the Russian.

As with any complex subject, the reasons why slums exist are many. Obviously, the basic reason is that the people living in them do not make a decent enough living to afford something better. As people flood into the cities from the country districts unequipped for city life, their earning capacity is low. The city is as ill prepared to receive them as they are ill prepared for the city.

When rural areas become depressed, from such causes as crop failures or overpopulation, then the surplus members of the farm families or the discouraged ones or the ambitious ones migrate to the cities. Boredom and the hope for a better life also lure the rural population. The slums of the city constantly ingest fresh victims.

We now know there is even a social law involved: the more the city raises its own standards, the more the rural people will migrate into it.

Housing projects to eliminate slum areas merely make the city more attractive to poverty-stricken farm area inhabitants.

The only sure way to reduce slum areas is the herculean task of raising the living standards of the farm regions to those of the city. This is approximately what has happened so fortuitously in Scandinavia and New Zealand. It would also be the situation in most of the United States were it not for depressed conditions in the South and Puerto Rico. Cities in the North and Middle West might be able to catch up with their slum problems were it not for the steady influx of Southern whites, Negroes and Puerto Ricans; how can city officials stem the spread of slum areas with housing projects when a never-ending additional supply of hungry mouths arrives each day by bus, jalopy and chartered plane?

If Americans with their wealth, aggressiveness and their tradition of striving always for a solution have not, so far, found it possible to eradicate slums, or even to lessen their area, consider the infinitely greater difficulty this problem poses for the officials of unwealthy Latin America, Asia and Africa.

This situation was well summarized by correspondent C. L. Sulzberger after a tour of the capital cities of Latin America:

> In most lands steadily mounting metropolitan attractions suck human energy away from a gradually stultifying countryside . . . As the cities boom the rural areas drift ever further behind. The cities get richer; the countryside gets relatively poorer; and the gap widens, with all its obvious and inherent political dangers . . . If Lima is attractive to [the rural Indians] in the form of [its slums'] waterless, lightless shacks scavenged by swine, how much more attractive will it be with some slum clearance . . . The necessity for slum clearance appeals to our decency and benevolence. But in overall planning does this really merit high priority? Cruel logic dictates a negative reply.

Slums are the result, not the cause, of national economic and social breakdown.

Slum clearance in the hungry nation is nonproductive. It does nothing to further the development of the nation's resources. It merely makes the few who are lucky enough to get apartments in the new buildings feel better. It is a luxury.

Worse, even those small sections that are improved at such great cost are immediately replaced by new slums in other marginal areas. Even if the entire city were rebuilt and the old one destroyed, new congeries of shacks, shanties and lean-tos would be procreating within a year as the rural population continued to swarm in from the discouraged countryside. Herein lies the fallacy that brings to naught the sums of money spent on slum clearance and urban development in the hungry countries.

Yet slum clearance is a wonderful, showy political gimmick. It can be *seen*. It can be completed within two years. It has propaganda impact. Hence the irresistible attraction it has to foreign aid officials.

The first three projects for which funds were allocated by the Inter-American Development Bank under President Kennedy's new Alliance for Progress were:

1. To Panama City, $7,600,000 to help finance 3000 houses in a low-cost housing development.
2. To Caracas, $12,000,000 to help finance 21,000 low-cost houses.

3. To El Salvador, $2,000,000 to help finance credits, mostly of medium- and long-term duration to small-scale farmers.

So $19,600,000 of the first $21,600,000 went to slum clearance in its various forms. Within the next eighteen months the United States had given Latin America $300,000,000 in loans, grants and investment guarantees for housing.

I hasten to emphasize I do not criticize the Alliance as a political gesture. The publicized goal of the Alliance, however, is not politics but to help Latin American countries to overcome their backwardness.

There is a second unfortunate result from improving slum areas. Despite chaotic conditions in these areas, the population is organized sufficiently, even in the most primitive cities, for public health departments to control epidemics. There are often such things as mobile TB units. Even the slum's grocery store, or bazaar equivalent, sells penicillin. You can walk behind the counter, drop your pants, lean over, and someone will give an injection. The city water is pure enough to form an adequate safeguard against contagious diseases. It may not be pure enough to protect germ-protected Americans from, to select some synonyms at random, Delhi Belly, Aztec Twostep, Djakarta Jumps and Turkey Trots, but the purity is sufficient to become a major cause of lower death rates and soaring population statistics.

Regardless of our compassion for the sick in these nations, high death rates are almost the only factor preventing even greater starvation and misery among the not-so-sick.

The rock-hard, dead-end trouble with slum clearance in the hungry countries is that the job is so enormous it cannot be done via housing projects. It is estimated that Latin America now has from 12,000,000 to 16,000,000 houses that do not provide "decent living." Yet William Vogt, a student on the subject, says it is likely that urban areas in Latin America will increase "more than 50 per cent in population within the next ten years."

In India for "the extra housing that will be needed by 1986 if the present rate of population growth continues, exclusive of rural areas or even of improvement of existing housing in such cities as Calcutta, the total investment required is estimated at approximately $25,000,-000,000," said Eugene R. Black when president of the World Bank.

"If you find the figure difficult to grasp . . . it is well over four times the total lent by the World Bank since it started business sixteen years ago."

So why waste time and—more important—waste precious government capital in this morass? Leave it to private money to build what it can at a profit, as was the procedure before government-financed housing came to dominate everyone's thinking.

The only way the government can itself alleviate the pressure of slums in these nations is to rehabilitate the rural areas sufficiently so the attraction to the cities is lessened. The problem is to keep them down on the farm and not let them migrate and congest the cities' unemployed and unemployable.

The overall goal is to make the rural areas more attractive to live in. A higher income through improved agriculture is the most effective method, but this is also the most complex and takes the longest time to achieve. Yet it can be realized if sufficient effort is given to solving the problem.

The technical problems of maximum land utilization and rural rehabilitation I take up later. Here, since slum clearance is a sociological matter, I offer some intangible sociological suggestions for making the rural population more aware of their own importance and for making life more pleasant, less stultifyingly boring. Cancellation of the smallest slum clearance project on the drawing boards will more than cover the cost of most of these.

Community centers in the villages. Not an antiseptic new hall. Just make certain the smelly *cantina* or teahouse or *serai* is enlarged to have a couple of pool tables and, horrors, a jukebox. Extending credit facilities to the *cantina* owner might be enough to set this up.

For the women provide a well with some stone flagging around it, or a cement laundry place with a roof, so they can congregate to wash the clothes rather than string out along the rocky banks of whatever rivulet passes for a stream.

Weekly movies. Not in an expensive new theater, but at any sort of open air spot. All that is needed is a $250 projector inside a waterproof shack on stilts. Come the rains and you will find the audience will be there just the same, each person with a poncho or umbrella or something over his head. Arranging these movies is not primarily a matter of expense but of organization. A highly successful advertising operation of an aspirin company in Central America was to send company trucks on regularly scheduled rounds to villages to show not advertising documentaries but regular commercial films.

Sports. I hesitate to suggest intervillage leagues, considering the explosive incidents that so often occur when sportive passions get out of control, such as those international imbroglios when rioters stone embassies and Ministers of Foreign Affairs sniffily break off diplomatic relations. Nevertheless, few things can develop a community's spirit so strongly as a good team. I once lived in Nebraska; it was then said the only thing that held together that spread-out state was the university's football team. I have already mentioned the village of Jicarito. There the greatest excitement was generated when the new soccer team acquired uniformly colored, tattered T-shirts. When the President of Honduras was due to visit the village, the team practiced running along the highway so they could act as escort for the presidential cortege.

New industries. When it has already been determined that a new factory is to be erected in a nation, the Minister of Resources should consider the advantage of helping it to locate not in the capital but in a rural area. This is one method of decreasing the number of persons migrating into the city. Perhaps special concessions in the form of transportation rebates or waiving of taxes can be established so that the new factory can compete with those based in the capital. But beware of the advocates of cottage, handicraft industries. Unless carefully thought out and organized in advance, these artificial efforts seem regularly to end in failure. A CARE official told me of the time he gave three sewing machines to a village in southeast Asia, brought in a sewing teacher, and established a class for making dresses and other clothing for sale in the capital city. The majority of the women, as soon as they learned this trade, immediately packed up, left the village and their husbands and moved to the city. They knew they would have a better living there as seamstresses. The poor CARE official had no idea how to solve this new problem of the wifeless village, for which he was, justly, blamed.

Taxes. A head tax might be levied on city inhabitants. This would at least emphasize to the prospective migrant to the city that living is more expensive there.

Better education. One attraction of the city for parents is the schooling available for children. It is urgent that rural education receive the same attention and monies spent in the cities.

Visits of high officials. One tangible way to create community spirit in a village and to develop a feeling of importance within rural areas is visits by the president of the country and by other high officials. It is also good politics, as many an expert executive has learned. Among those who have demonstrated that tumultuous city folk can be downgraded if the rural areas are calm, Magsaysay and Nasser come to mind. The only examples I recall that ignored the agricultural population in favor of city workers were Stalin and Perón.

Japan was able to recover from the emotional and physical chaos, after the war, largely as a result of rural stability.

Ubico, the dictator of Guatemala (1931–1944), based his political support on the rural districts and almost completely ignored the citizens of the capital. At every opportunity he made trips through the countryside. It is still a legend in many places how Ubico would suddenly appear, where roads had never been built, riding his own motorcycle up the rocky mountain trails.

Lázaro Cárdenas, President of Mexico (1934–1940), developed and maintained his firm political support by countless, and I mean countless, hard physical trips into the most difficult, remote areas.

My first post in the Foreign Service was in Mexico City during the regime of Cárdenas. My ambassador, Josephus Daniels, elderly and seemingly frail, often accompanied Cárdenas on these trips and, equally important, often went off on similar rough trips by himself. Usually Mrs. Daniels, also elderly and also seemingly frail, went with him. They did more traveling around the country than the rest of the embassy staff put together. During those difficult political days of the Mexican expropriation of the oil companies' properties and the land-reform cutting up of large *haciendas*, including American-owned ones, this steady traveling of the ambassador was a major factor in lessening the anti-American feeling and keeping it within manageable limits.

I emphasize these points because I want to accentuate the importance of the agricultural areas in the political life of most backward countries. Remember: these countries are *agricultural*. Usually, the capital is the only city.

The government's primary goal should be to give the rural areas a feeling of importance, a feeling of pride, that they do indeed form the backbone of the nation. Once such a feeling has developed, the pull of the city for the rural population may be less magnetic.

The poor in the cities are concentrated in their slum areas. They are easily aroused by orators and easily led into demonstrations and rioting. They are a physical, tangible force whether active in the streets or latent under police controls. They can needle a government on a violent, temporary basis, whereas the scattered inhabitants of the countryside do so only on an intangible long-range basis. But any political scientist will affirm that almost always it is the long-range influence that prevails.

Also pertinent: the poverty in the rural areas makes less impression on the casual visitor. It is not readily seen. Often it is difficult to determine, as one passes by, whether a farm is relatively prosperous or relatively impoverished. In the cities there is no doubt. Slums make dramatic magazine photos; farms are merely picturesque.

In the policy-making rooms of Washington, voices are constantly raised that this or that type of aid project, especially slum clearance, is vital because it will gain the good will of the urban masses. The voices argue that this is where lies the fulcrum that controls the country's political life. Usually, it would seem, these voices are frightened by the headlines of the latest riots of university students. It would also seem that this is one more example of American officials extending their own modern urban background into the quite different environment of the hungry nations. City bosses and ward machines and labor unions, even of the currently fashionable "clean-cut" type of the United States, do not dominate the life in the hungry nations. These nations are *agricultural.*

The chairman of the United States Senate Subcommittee on Housing, Senator John Sparkman of Alabama, has announced that what foreign aid needs most is more housing projects. When he appointed a team of thirty-six "scholars and housing experts" to prepare a study of international housing, he said, "Our foreign aid economists should modernize their thinking, recognize housing not only for its social purpose but for its economic impact, which, in the United States, has made it the fifth largest industry." (Note here, again, the unquenchable dogma that what has been good for good old U.S.A. is dandy for everybody else.) "Too often in the past housing has been relegated to the bottom of the totem pole in our foreign assistance programs. I believe the American people would have a great deal more sympathy with our foreign aid program if it were oriented more toward direct assistance to the people, such as housing."

Senator Sparkman made no mention specifically of slum clearance, but neither did he talk of adobe and/or palm-thatched huts complete with dirt floors. In the hungry nations any house with solid walls, a tile floor, a roof that does not leak, and a total of three rooms—one a kitchen lean-to, one a parlor, and one a bedroom for momma, poppa and the seven kids—any house of such grandeur is strictly for the bourgeoisie. And yet is this not what the Senator's team of "scholars and housing experts" and the slum clearance planners are aiming at?

A community development worker who had spent most of his life working for the Friends Service Committee told me that in tropical America and Africa tile or cement floors are not as sanitary as packed dirt floors. For families that do not look on cleanliness as does an American or European middle-class family, a dirt floor allows the children's urine to filter through. The puddles left by children, dogs and other animals do not persist. Also, in his view, a roof that is good enough to protect the inhabitants from rain is all that is needed to make any house classifiable as acceptable. Would such a house fit Senator Sparkman's standards?

I would not be so foolish as to say people prefer to live in hovels. However, it is also foolish to try to synchronize the feelings of a middle-class American concerning life in a hut with the feelings of an Indian who has known nothing else all his life.

Sociologists have noted that most American welfare workers and ministers come from our middle-class families where cleanliness is next to godliness, where it is not *nice* to fight with your fists, and where it is not *nice* for parents to sleep in the same bedroom with the children. Yet these are the welfare workers who deal with slum families where reverse standards are accepted. One result is that legislation (in our case, foreign aid allocations for slum clearance) is oriented towards putting the straitlaced *mores* of the American middle classes onto the alien lower classes. These lower classes, nevertheless, have their own firmly entrenched, strictly moralistic standards. Who is to say which set of morals is the better?

An article in the *National Geographic Magazine* included an attractive photograph of a new government housing project in Athens. The caption said that these neat, balconied apartments were sold on twenty-year terms to urban dwellers at $0.13 per month (thirteen cents, just to make sure you do not misread the figure).

Who would not rush to the city pell-mell when he has the chance to buy a fine new apartment for $33.20, including interest, in payments spread over twenty years! Wouldn't you?

When the steady migration into the cities is lessened, the dead-hand pressure of the slums on the economic life of the cities will decrease. Agitation for slum clearance should similarly decrease, but the slums themselves will remain—indefinitely.

Meanwhile, leave the construction of new housing to private capital so at least some money profit can come from it. Channel government capital (tax income) into development work that will redound to the national advantage.

THE DIRTY
ANIMAL—MAN!*

JOSEPH L. MYLER

Joseph L. Myler is one of America's top science writers. Currently he is senior editor for United Press International.

 In his beautifully written series of essays entitled The Dirty Animal—Man!, you will not only be impressed by his articulate use of the language, but you will feel a certain inner uneasiness about the environmental man-made mess.

WATER POLLUTION

Man is poisoning his world.

He has been labelled, with strong justification, "the dirty animal." He has managed to make his rivers rotten. He has transformed green pastures into deserts. He has clogged the air with chemicals which menace health and dust which is changing the climate. He is a menace to himself and other species.

He has turned large areas of his world into junk heaps, piled high or layered deep with indestructible cans or plastic containers. Americans alone discard more than a billion tons of solid waste a year and the total is growing.

Man is beginning to face up to the problem, but only slowly and against great obstacles because of government and industrial considerations.

Unless he is willing to spend billions upon billions to undo what he has done—and perhaps even change some of his basic ways—he really may be gasping for breath in a few decades.

In the developing nations, nearly a billion people get their water from unsanitary sources and half of them get sick every year as a result. Even in the United States, half the people depend on water supplies which don't meet federal standards or are of unknown quality.

Rivers of so-called developed nations have been turned into sewers of civilization to get rid of unwanted industrial wastes. The oceans are being contaminated with agricultural poisons which drain into streams and are carried away to the seas.

By exhausting warm water from our power cooling plants into the ocean, we are threatening marine life. All along we have drained the priceless topsoil of our fields into silting rivers. We have denuded many of our forests.

Millions of workers are exposed to potentially dangerous concentrations of dust, fumes, gases and vapors. No one knows what the noise generated by modern machines and cities is doing to man's nervous system.

As J. George Harrar of the Rockefeller Foundation has said, "Man himself is the greatest threat to his environment. . . . We have now successfully begun to contaminate what we have not yet destroyed."

None of this happened overnight.

Once the oceans were thought to be endless, the land infinite and the atmosphere limitless. Now man's survival is known to depend on how he husbands a relatively thin layer of soil, water and air tightly wrapped around our planet's surface.

Nature with its wind and water erosion and climatic changes has been altering the environment for millions of years. But the possibility that one species might make the world uninhabitable did not arise until 8,000 years ago when the hunter, who simply ranged the land in search of food, evolved into the farmer who plowed and uprooted it.

Next came the city and then the industrial society, which multiplied the threat many times over.

Dr. Barry Commoner, Director of the Center for the Biology of Natural Systems at Washington University, said, "Modern technology has so strained the web of the processes in the living environment at its most vulnerable points that there is little leeway left in the system.

"Unless we begin to match our technological power with a deeper understanding of the balance of nature, we run the risk of destroying this planet as a suitable place for human habitation."

Consider what man has done to one indispensable element of our biosphere—water.

Water in a sense is the most precious stuff on this planet.

Yet we waste it, we pollute it, we threaten the existence of irreplaceable underground reserves which took Nature thousands of years to establish, we destroy the beauty and the life of once sparkling streams and deep blue lakes.

What is so precious about water—that cheap fluid most of us in this country can get in any amount just by turning a tap?

Throughout human history water has been the greater limiter. No civilization has ever risen without a plenitude of water. When water runs out, or becomes unusable, civilizations die.

Men have killed each other for water, whether at some isolated spring in the 19th century American West or in ancient Mesopotamia where human beings warred for control of the Tigris and Euphrates.

Water is one of the reasons for today's bloody rivalry between the Israelis and the Arabs.

The high standard of living in the United States and other affluent nations of the modern world depends on fresh water—lots of it.

Americans use about 310 billion gallons of water a day on the average for public supplies, commerce and industry, irrigation, and rural domestic and livestock needs. On a per capita basis, this is 1,600 gallons a day.

Of the annually renewable water supplies available to the United

States, about 1.2 trillion gallons a day enter the streamflow from surface and underground sources.

This amount, 1.2 trillion gallons a day, constitutes the nation's ultimate water resource—for homes, industry, irrigation, recreation.

Properly managed, it can be used and reused before release to the oceans. Only a tiny amount is "consumed" in the sense that it is converted into other forms, such as chemical products, or removed as a resource by being turned into vapor.

So, the United States is water rich. With all that magnificent streamflow it can never become thirsty. Or can it?

For one thing, the figures are all in averages. The blessing of fresh water from the sky ranges from less than an inch a year in some parched regions of the Southwest to more than 200 inches in the Pacific Northwest and parts of Hawaii.

For another, populations and the demand for water are rising faster than man's means for making his water resources available wherever needed for human use.

The world population is expected to double to nearly 7 billion by 2000. Says Dr. Raymond L. Nace, Research Hydrologist of the U.S. Geological Survey:

"The problem is not whether water supplies are running out, but whether people are outrunning the supplies. Water supplies have finite limits, but the demands of people on the supplies have no known limit."

Unless he gives up piecemeal, temporary solutions to local water problems and concerns himself with the long-term global problems, man will be in trouble. For that matter, he already is.

Take pollution. To list the specific pollutants which man dumps into his water supply would take many pages.

They range from raw sewage to chemical fertilizers and animal dung, from acids and poisons generated by industry to silts and salts drained from strip mines, city streets and farmlands, from crankcase oils and detergents to disease carrying bacteria, from herbicides and pesticides to radioactive contaminants from mines and atomic plants.

Congress has enacted laws to control water pollution and is studying new ones. But the pollutant load is steadily increasing, and some of the problems involved seem almost too difficult to be solved by legislation alone.

Listen again to Nace of the Geological Survey:

"Out of its total potentially controllable liquid assets the United States uses 95 per cent chiefly as a conveyor belt on which to send waste products out to sea.

"The major use of free-running water in industrial nations is not industry, as published statistics seem to show, but waste disposal. Our rivers are open sewers."

Others have said it even more starkly.

Dr. Glenn T. Seaborg, Chairman of the Atomic Energy Commission, says all 22 river systems in the United States will be "biologically dead" by the end of this century if pollution continues at present rates.

According to Hollis R. Williams of the Agriculture Department's Soil Conservation Service, "The pollution of the living waters of the United States is one of the great shames of our time."

According to former President Lyndon B. Johnson, "The clear, fresh waters that were our national heritage have become the dumping grounds for garbage and filth. They poison our fish, they breed disease, they despoil our landscapes."

These pollutants also are killing some of our lakes. Nutrients from wastes or farm fertilizers have created "algal blooms" which result in depletion of oxygen in the water, destroy fish, and set the stage for ultimate transformation of a lake into a marsh and eventually a meadow.

Lake Erie may already be doomed by this cycle. Lake Michigan is in danger.

According to the recent report of the Marine Science Commission, man has created a "devil's brew of pollution" which constitutes "a growing national disgrace."

How serious is all this in the world scheme? Dr. LaMont C. Cole of Cornell University has warned that mankind seems bent upon his own extinction.

THE CRISIS ON THE HORIZON

Without water there could be no life of any kind on earth. In a sense water is even more precious than oxygen, the "gas of life." For without water there would be no green plants, and green plants supply the oxygen in the air we breathe.

Scientists believe life on earth originated in the primitive seas long before there was more than a trace of oxygen in the atmosphere. Oxygen, and life's dependence on it, appeared only after the evolution of plants.

The blood of animals, including man, still is a salty solution similar to sea water. The sea still surges in the circulation systems of land as well as marine creatures. Most living things are mainly water.

The sea is at once the supplier of fresh water to the land and of oxygen to the air.

More than 70 per cent of our oxygen supply, according to Cornell's Dr. LaMont Cole, comes from microscopic green plants in the sea which, like the plants of land, consume carbon dioxide with the help of solar energy and cast off oxygen as a waste product.

With his bulldozers and concrete and asphalt city-building, road-building, urbanizing man has destroyed oxygen-producing vegetation over large continental areas.

However, enough plant life, the phytoplankton, remains in the oceans to keep the oxygen content of the atmosphere fairly stable at about 20 per cent. But man is polluting even the oceans, with what consequences he does not know.

In the solar system, at least, earth appears to be uniquely blessed with water in great quantities. Only in the case of Mars does there appear to be any faint possibility that another planet of the sun's family has or even has had any liquid water at all.

On earth there is a prodigious amount of water—326 million cubic miles of it. Of this hard-to-conceive quantity, 317 million cubic miles are in the seas which cover 71 per cent of the globe.

Most of the rest consists of "frozen assets" of fresh water locked up in glaciers and the polar icecaps.

Man, of course, is primarily concerned with available fresh water, the stuff he can drink or moisten his yards and crops with, or use in cooking, washing, and industry, or as a medium for harboring trout and other fish which it is fun to catch.

For recreation men do, of course, swarm to the sea beaches, and the estuaries, and release their tensions in many salt water sports—surf swimming and fishing, scuba diving, sailing, and lolling on the sand in the sun. They transport most of their goods in world commerce upon the salty oceans.

But the sea's great gift to man is fresh water. The sun annually distills (evaporates) 80,000 cubic miles of fresh water from the oceans and 15,000 cubic miles from the land.

At all times about 95,000 cubic miles of water are moving between earth and sky. What goes up subsequently comes down. This, crudely put, is the hydrological, or water, cycle.

This water, as rain, snow, hail, or sleet, comes down all over the world. Most of it falls back into the oceans. But a lot of it falls on the land. The United States gets about 30 inches a year, or 4.3 trillion gallons a day. Roughly 70 per cent of this is sent back up into the air as vapor. This includes the water used by plants.

It seems silly to talk of polluting the ocean. But it is happening. DDT has been found in marine creatures everywhere. And if the plant of the ocean is jeopardized, so is the oxygen supply on which all life depends.

The Torrey Canyon oil spill of 1967 and more recent ones, including the calamity off Santa Barbara, Calif., were disasters. Animal and bird life in the spoiled areas may never be the same. Perhaps, just perhaps, these calamities were strictly local.

In any event, they might have been worse, given man's capacity for unintentional destruction. Suppose the Torrey Canyon had been loaded not with oil but with herbicides.

Cole asks the question: Would photosynthesis, the process by which plants produce oxygen, have been wiped out in the North Sea?

A few such accidents could leave man gasping not in a matter of generations, Cole suggests, but in a matter of years.

An alarmist notion? Possibly. But those who have looked hardest at what man has done and is doing to his environment have come to expect the worst.

Some authorities hold that for the United States, at least, there is no water crisis. Says the National Academy of Sciences, "There is no nationwide shortage and no imminent danger of one."

It goes on to say, however, that "There are serious regional shortages of usable water, many of which are becoming critical because of short-sighted planning or pollution of fresh-water supplies."

Nace, of the Geological Survey, agrees that "No human crisis centering around water exists today." "But," he adds, "one is visible on the horizon."

He continues: "We have surely been living in a dream world of water abundance at prices cheaper than we pay for common dirt."

Former Secretary of Agriculture Orville L. Freeman fears that "Future generations will judge most harshly a race of men that had all the technical knowledge, all the resources they needed to provide a clean water, air, and land, but lacked the will to do so."

And, says Freeman, "We are facing an environmental crisis. It affects every one of the basic elements of the biosphere—air, earth, and water, and every one of us."

Nace recently pleaded for preservation of a resource which he said is "perhaps the most valuable" humanity possesses. This is what the hydrologists call ground water. It has been stored by nature over the millennia in subterranean "aquifers" consisting of porous rock, gravel, sand, and sediments.

According to Nace, 97 per cent of all fresh liquid water on the continents is contained in aquifers which hold "many times more water than can be stored in all the surface reservoirs that will ever be built" by man. They are "buried treasure."

In arid regions they constitute the chief source of water. This nation, Nace said, need never run out of fresh water if it cooperates with Nature to maintain its aquifers.

But "if these resources receive the same reckless treatment that surface waters have, we will destroy the usefulness of our only real national water resource."

Ground water supplies are menaced in many ways. They can be killed by overpumping which results in subsidence and compaction of subsurface materials to the point where they become impervious and hence useless for water storage.

They also can be made unfit for use by pollution. Encroachment of salt water into pumped coastal aquifers is one source of pollution. Septic tanks do their bit. Another source is the growing practice of underground disposal of industrial wastes.

"To sweep our wastes underground now," Nace says, "may create an unsolvable problem for the future."

One of man's new and weird pollutants is simply warm water. Most of the water taken by industry and cities from streams is used for cooling and then poured back.

According to scientists, many forms of aquatic animal and plant life are threatened by the great tonnage of heated water from power plants, nuclear or coal-fired, which is being spewed into rivers, lakes and coastal waters.

If the problem is one for the present, it is even more so for the future.

A study reported by the Geological Survey showed that in 60 of the undeveloped countries of Africa, Asia, and Latin America 90 per cent of the population depends on water supplies "that are inadequate or unsafe."

The shortage in all countries, according to Nace, is not of water but of waterworks to make the available water usable.

The United States, with tough regional water problems of its own, is trying to help less fortunate nations with theirs. In 1967 it created the Office of Water for Peace in the State Department. This agency is concerned with a host of projects ranging in scope from provision of drinkable water on a local scale to large river-basin development programs.

As part of the Water for Peace endeavor the United States is spending about $400 million a year in many countries to build waterworks designed to supply both household and industrial needs.

This is something that needed to be done. No nation can mature without an abundance of water. But does anybody imagine this or anything else projected will satisfy the needs of the seven billion human beings who will populate the earth by 2000 A.D. if current forecasts come true?

Nace is appalled by these population predictions in view of the inability of men "to control either nature or themselves."

"Imagine what the pollution load on water supplies could be with that many people around! Especially if the 'advanced' countries succeed in teaching the retarded ones all of their technologically ingenious ways for adding new and weird pollutants to the environment."

COMING TO TERMS WITH NATURE

About 60 per cent of the earth's land area is dry or downright arid, incapable of supporting agriculture. It now accommodates only about 5 per cent of the world's people. But if the population doubles in the next few decades as predicted, these now largely empty spaces must be transformed into "living spaces."

The deserts must be made to bloom again.

Can this be done? Has mankind learned anything from the mistakes of the past, some of which actually created deserts where none existed?

In the United States we have stripped vast acreages of trees and sod. We have built great cities and greater suburbs. The result has been to speed the runoff of water and to increase the frequency of flash flooding.

Running water is power, and we use it to manufacture electricity and create huge reservoirs that are a blessing to farmers and a joy to fishermen and others seeking solace in and on the water. Not the least thing to be said of these manmade lakes is that they are beautiful.

But we also have created silting and erosion problems, and have destroyed much of the thin layer of topsoil which is the sole source of our agricultural wealth.

Not all floods, of course, are to be blamed on man. But in many regions, notably Southern California, they have been aggravated by what man has done to the land.

Every year about 75,000 Americans are driven from their homes by floods. Damage is about $1 billion. As the Environmental Science Services Administration says, "Floods are also great wasters of water, and water is a priceless national resource."

To save and increase water resources and prevent or at least mitigate the effects of floods, this country has spent many billions. It can hardly be doubted that this money has been well spent.

But are we really tackling our water problems in the best way—for us and our descendants? There are those who fear we are not.

In the Far West we pump out our ground water for decades before we learn we must return unconsumed surplus water to these precious reservoirs to keep them alive. So the central valley of California is gradually sinking and killing some of its subsurface water resources.

In the southern high plains of Texas and New Mexico we are depleting underground water stores which would take Nature many centuries to replenish if they were exhausted.

Dr. Barry Commoner, director of the Center for the Biology of Natural Systems at Washington University, has said mankind is suffering from an "environmental disease" created by technology.

"We are unwitting victims of environmental pollution," says Commoner, "For most of the technological affronts to the environment were made not out of greed but ignorance. . . . In each case the new technology was brought into use before the ultimate hazards were known. We have been quick to reap the benefits and slow to comprehend the costs."

According to hydrologist Raymond Nace of the Geological Survey, "The history of human effort to control Nature is a history of continually having to combat the unwanted consequences of these efforts."

But man does have to try to solve his water problems. He has learned to recharge his aquifers with what otherwise would be water wasted as runoff.

He also is learning to conserve water resources he used to throw away.

Santee, a town near San Diego, has provided an example of what can be done to get the most out of scarce water supplies. By treating and filtering sewage, it is restoring a million gallons of water a day to drinking quality.

In Los Angeles County one purification plant reclaims 10,000 acre-feet of water a year out of a trunk sewer.

The Bethlehem Steel Co. takes the entire sewage of the city of Baltimore and, after treatment, uses it for industrial purposes. The total is more than 125 million gallons a day.

Forgetting the billions of future inhabitants of this planet for the moment, present emergencies call for immediate solutions.

Desalination of seawater is one of these. The Atomic Energy Commission has a grand thought for changing desert coastlands of the world into latter-day oases.

Build huge nuclear plants capable of producing a million killowatts of electricity and at the same time, with nuclear heat, desalting 400 million gallons of water a day.

Such a system, nicknamed a "nuplex," could produce enough power, water, and fertilizer to raise more than a billion pounds of grain a year at the site, and enough extra fertilizer to grow enough food to feed "tens of millions of people annually."

Such "agro-industrial complexes," according to the AEC, might make fit for human habitation the "one-third of the world's land" which is now "virtually unoccupied—in Australia, India, Mexico, the Middle East, Peru, and the United States."

This, like NAWAPA, is still only an idea. The AEC and a group of California backers had planned the world's first nuclear power–desalting plant on Bolsa Island, a manmade spit of land off the California coast near Los Angeles. It would generate 1.8 million killowatts of power and produce 150 million gallons of fresh water daily, for an investment of $765 million.

The cost turned out to be too great, the sights were lowered, and the project is now in abeyance. No nuclear power and desalting plant has yet been built.

But desalting by other means has been undertaken. There are about 650 plants in the world capable of producing 250 million gallons of "demineralized" water a day. The biggest of them supplies Key West, Fla., with 2.6 million gallons of freshened sea water a day at prices lower than Key West used to pay for water brought by aqueduct from the Florida mainland.

Others operate in such widely separated places as Kuwait, at the head of the Persian Gulf, and Malta in the Mediterranean.

Important as these, and the far bigger plants envisaged by the AEC "nuplexes," may be on a local or regional basis, their output is a mere drop in the bucket compared to the global water supply.

If you multiplied the world's desalting plant capacity many thousands of times, according to hydrologist Nace, the "contribution to the total water supply still would be extremely small."

In other words, desalting of sea water is by no means the panacea man has been looking for. He has to look somewhere else. Even if he could convert the sea from salt to fresh water, what would he do with the salt? He would be back where he started.

Where else can he look? Pending the long-range solutions, when man knows more about the complexities of the hydrological cycle, what can he do?

There being no global water shortage—there's as much water in the sea as there ever was—it is necessary to do whatever can be done to ameliorate local shortages, now.

The Federal Council for Science and Technology in 1966 proposed a 10-year program for water resources research. It called for stepped-up research, not overlooking the field of "far-out ideas."

A decade ago a California oceanographer noted that some big icebergs from Antarctica drift far enough north to come within capture distance of interested persons above the Equator.

He suggested that bergs might be towed to anchorage off Los Angeles, say, and be made to give up their locked-in fresh water for the uses of man. The notion of a crackpot?

The Federal Council warned against such unthinking dismissal of possibly good ideas. It recommended investigation "to ensure that worthwhile concepts are not overlooked."

In the meantime there is Nature to think about before embarking on any projects affecting truly large parts of her atmospheric, aquatic, and continental domains.

Listen again to Nace:

"Man acts for his own purposes and Nature reacts according to her immutable laws. Nature, a philosopher has said, is neither friendly nor inimical. She is merely implacable . . . we had best come to terms with her . . . .

"Until we do so, our so-called conservation practices are likely to be mere tinkerings with the landscape."

AIR POLLUTION

A nebulous veil hangs ominously over the earth. Man put it there. It is polluted air, loosely called smog.

Its presence proves that man has the technical capacity to suffocate himself, and other living things on this planet.

There may be time to rend the veil, or at least to keep it from reaching the dimensions of disaster.

In many places the veil still is imperceptible to the senses. In others, its presence often is painfully apparent.

When the eyes smart and water, when breathing is a throat-burning ordeal you know—even if the sun is bright and the sky lovely—that smog is all around you.

The atmosphere is vast. For billions of years it possessed the ability to keep itself fairly clean, clean enough to sustain life on earth.

But man has established himself as superior to Nature, if only in a self-defeating sense. He can foul his environment faster than Nature can clean it up.

Man has polluted his rivers and lakes and soil, and even the seas.

He has produced a crescendo of noise and clatter in his factories and offices—yes, and in his places of amusement—which not only impair efficiency but actually threaten health.

He has strewn his highway and byways, his streams and his beaches with the ugly litter of civilization. What to do with solid waste has become a gigantic puzzle with ill-fitting pieces.

The atmosphere which wraps the planet in a gaseous film no thicker in proportion than an apple skin has not escaped.

In April, 1968, J. George Harrar, president of the Rockefeller Foundation, addressed the annual meeting of the National Academy of Sciences. Outside azalea bushes were flaming all over Washington, robins were chirping, all seemed right with the world. But inside the Academy Harrar was saying:

"And now it is our air envelope that is endangered. The outpouring of the byproducts of modern industrialization has reached dimensions with which we are at present unable to cope.

"Incinerators, industrial and power plants, automobiles, and many other elements combine to produce a complex pollution pattern. As all nations increasingly industrialize, and as their cities burgeon, the possibility of eventual suffocation as the result of this pollution becomes a very real threat."

There appears to be no doubt that air pollution is a serious menace to public health. There appears to be no doubt that air pollution, unchecked, can change the world's climate, probably for the worse. If the polluted air doesn't choke us, it may trigger climatic events such as melting of the polar ice caps which would drown our greatest cities. Conversely it may hasten the return of the Ice Age.

But there still is, say some experts, time to prevent these calamities. How much time? The Committee on Pollution of The National Academy of Sciences reported:

"Air pollution is increasing faster than our population increases."

The world population is expected to double by 2000 A.D., 31 years hence.

Another academy group, the Committee on Atmosperic Sciences, estimated that "we have, at best, but one or two generations" in which to acquire understanding of the problem and do something about it.

There are those who wonder if the effects of pollution have not already become irreversible. If they have, we live on borrowed time, and the question is how much time there is left to borrow.

Even if smog were not a threat to health and the future of the race, it is dirty and costly. Damage done by air pollution to buildings, trees, crops, livestock, materials and works of art has been estimated at $13 billion a year in the United States alone.

Even before man the atmosphere was polluted. Volcanoes have thrown dust and gases into the air from time immemorial. Other natural pollutants have been dust from wind-eroded soil, salt particles from the sea, methane from marshes, hydrogen sulfide from rotting organic matter, radioactive materials from the earth's crust, pollen, spores, and airborne bacteria.

With these natural pollutants Nature could cope. The atmosphere starts at the earth's surface and rises hundreds of miles until it fades into space. It weighs a stupendous 5,700 trillion tons.

About 70 per cent of this great ocean of air—thin though it is—exists in a region about six miles high. This is in the troposphere where the winds blow and the rains and snows fall and the hurricanes and tornadoes roar.

This is where the pollution, man-made or natural, tends to accumulate on the calm days. This is where Nature, with wind and rain and snow does what it can to disperse or precipitate the pollutants which otherwise would clog it.

The atmosphere normally consists of about 78 per cent nitrogen, a relatively inert gas; 21 per cent oxygen, which keeps us alive; three-hundredths of 1 per cent carbon dioxide, the prime food of plants, and smaller amounts of other gases.

Now comes man with the fumes and particles released by his fires and engines and the dusts blown from the deserts he has made and the pesticides he spreads to the four winds and the solid and liquid bits of waste he dumps into the air. He threatens not only to defeat the atmosphere's self-cleansing capacity but also to change drastically its fundamental composition.

In this country we pour about 143 million tons of pollutants into the air in a single year. In thus using the atmosphere as a sort of garbage dump we unwittingly make co-conspirators of the weather and the sun.

Fog grabs man's noxious or toxic aerosols and other particulates and concentrates them in stifling pockets. It converts sulfur dioxide, a product of burning coal or oil, into sulfuric acid. Fog and sulfur oxides have figured in all the great air pollution disasters of the past.

Even the sun, the mainstay of earthly life, compounds man's felony. It plays on the hydrocarbons and nitrogen oxides from internal combustion engines and transforms them into irritating or sickening components of "photochemical smog" such as ozone, peroxyacyl nitrate (PAN), and formaldehyde.

On occasion the weather creates conditions that intensify the evils already present in the atmosphere. A layer of warm air slides above the cooler surface air. This is a "temperature inversion." It keeps the dirtiest air from rising and dispersing. It traps atmospheric poisons emitted by the works of man.

Such inversions are common. Stagnant air frequently covers huge areas of the United States, particularly in fall and winter. Such times, variously called "squaw winter" or "Indian summer," can be beautiful beyond compare. They also can be dangerous for those who suffer from heart disease or the manifold respiratory maladies, asthma to emphysema, which are aggravated by smog.

SMOG AS A KILLER

"Smog" is a word coined centuries ago in Britain to describe the ugly mixture of smoke and fog that sooted the countryside for miles beyond the environs of coal-burning London.

As the word is used today, smog means something much more complex than a mere mixture of smoke and fog. It is the photochemical variety that plagues Los Angeles, for example, on 200 or more days a year. Los Angeles is hemmed by a bowl of mountains which interferes with air circulation and helps to maintain stagnant inversions.

One result has been that Los Angeles leads the nation in pioneering efforts to do something about air pollution. It has at least kept the problem from getting worse for the time being.

What troubles Los Angeles and practically every other large city in the world is a chronic condition which, however unhealthy it may be, still lies somewhat this side of disaster.

It was the acute "episodes" that forced upon the world some awareness of what industrial man can do to the atmosphere and himself.

As the National Tuberculosis and Respiratory Disease Association said in a recent publication:

"It was probably the shock of the notorious air pollution disasters . . . that first stripped the smokestack of its glory . . . and turned what was a monument to progress into a gravestone for the dead."

Some disasters:

Meuse River valley, Belgium. Belgium was blanketed by a thick, cold fog in the first week of December, 1930. The heavily industrialized, 15-mile-long Meuse River valley was trapped under an inversion layer of stagnant air. Pollutants accumulated. In a few days thousands became ill; 60 died.

Donora, Pa. An inversion covered a large part of the northeastern United States in October, 1948. In the Monongahela River valley lay Donora, a town of 14,000 crammed with industries. Nearly half of the inhabitants, 6,000, got sick. And 20 died.

London. For centuries Londoners had heated their homes with soft coal in fireplaces. London fogs were both notorious and commonplace. But in December, 1952, there occurred a fog which surpassed others of the past. A five-day spell of stagnant air killed 4,000 persons.

New York. This city's pollution disasters might have gone unnoticed if public health scientists had not studied the death records. They concluded that New York had experienced air pollution "episodes" in 1953, 1962, 1963, and 1966. In 1963, they found, 405 persons died because of poisoned air.

The chief victims in each disaster were the elderly with ailments of the heart and lungs. Weather conditions helped the polluted air to do its deadly work.

Then, as the National Tuberculosis Association points out, there have been the "industrial accidents." One foggy morning in November, 1950, in Poza Rica, Mexico, an accident at a sulfur factory resulted in a spill of hydrogen sulfide. In half an hour enough of this poison, which gives rotten eggs their loathsome stench, had been spewed into the air to sicken 320 persons and kill 22.

But there is more to air pollution than acute, sudden disaster. There is chronic daily pollution and the chronic daily damage.

The major pollutants are carbon monoxide, sulfur oxides, hydrocarbons, nitrogen oxides, particulate matter (everything from aerosols to soot), and what the official tables call "miscellaneous other."

This last category includes lead (from auto fuel), fluorides, beryllium, arsenic, asbestos, many other chemicals, and a host of pesticides, herbicides, and fungicides.

The major pollutants are autos and trucks and jet airplanes, power plants, space heating, refuse disposal (dumps and incinerators), and various industries—pulp and paper mills, iron and steel mills, oil refineries, smelters, and chemical plants.

Together in the United States they throw 143 million tons a year of "waste" into the air, and (say the experts) the automobile is the greatest villain of all.

According to a recent Senate report the auto accounts for at least 60 per cent of total U.S. air pollution, 85 per cent of pollution in the big urban areas, 90 per cent of all carbon monixide pollution.

In all, the auto exhausts more than 90 million tons of pollutants into the air each year—twice as much as any other fouler of the atmosphere.

If none of the fuel burned by man contained impurities (such as sulfur in coal and oil) and if all of it underwent complete combustion, the byproducts would be simply water and carbon dioxide, neither of them harmful within limits and neither listed as contaminants.

According to Dr. LaMont C. Cole of Cornell University, half of all the fuel ever burned by man has been burned in the past 50 years. Since such fuels as coal and petroleum are "a non-renewable resource," this means that man is operating an exploitive economy that "will destroy itself if continued long enough."

Moreover, this dumping of carbon dioxide into the air (six billion tons a year) threatens to change the climate and may indeed already have changed it. It also endangers the atmospheric oxygen balance which has sustained life on earth since Nature invented photosynthesis, the process by which, with the help of sunlight, plants take up carbon dioxide and liberate oxygen.

Carbon dioxide is transparent to sunlight but tends to absorb heat radiated back toward space from earth. This is what meteorologists call the "greenhouse." It has been estimated that the planet's average temperature would be 20 degrees cooler if there were no carbon dioxide in the atmosphere.

It also has been estimated that man with his burning has increased the air's carbon dioxide content 10 to 15 per cent in the past century and will have increased it 25 per cent by the year 2000.

A worldwide warming trend was noted from 1900 to 1950. Among the effects was to move the crop line 50 to 100 miles north on the Canadian prairies. Mocking birds, once common only in the South, extended their range and their sleep-shattering song to New York.

But this trend apparently has been halted and even reversed by the other things human beings are doing to the air. Dust, smoke, and other particulates, particularly over the world's cities, are cooling the planet by reflecting sunlight away from the ground.

Doubling atmospheric carbon dioxide could increase the earth's average surface temperature nearly seven degrees. But only a 25 per cent increase in turbity (dust, smoke, liquid particles) could lower it by the same amount.

Dr. Reid A. Bryson, University of Wisconsin climatologist, believes the veil of pollution hanging over the world already has changed its climate.

DDT dust from farm fields has been carried by winds to all corners of the earth. Soviet cities have increased smokiness over the Caucasus 19-fold since 1930. Turbidity of the air over Washington, D.C., has gone up 57 per cent in recent years.

Over Switzerland, Bryson says, it has jumped 88 per cent. In the decade between 1957 and 1967 there was a 30 per cent increase of dustiness over the Pacific Ocean. Smoky days in Chicago rose from 20 a year before 1930 to 320 in 1943.

A blue haze, probably from agricultural burning, hangs over Brazil, southeast Asia, and central Africa. A brown haze of dust from soils made barren by man broods over much of Africa, Arabia, India, Pakistan, and China.

The sometimes-not-so-nebulous veil is global in extent. The jet airplane is helping to thicken it. According to Professor W. Frank Blair of the University of Texas at Austin, the city of Dallas is visible from transport planes many miles away because, in part, of the exhausts of jet aircraft.

Jets discharge tons of particulate matter into the air every day. Their contrails, says Bryson of Wisconsin, trigger formation of high-altitude cirrus clouds which tend to alter climate by reflecting sunlight back to space.

At the time when much is being said about intentional modification of the weather (which is a long way off) climatologists are worried most by inadvertent changes. This is because they can't really guess the ultimate results.

Suppose the greenhouse effect kept piling up. The ice caps would melt and the sea level would rise, perhaps as much as 400 feet. This would drown not only coastal communities but vast low-lying inland areas. It also might have the effect, since the increase in water vapor would make snowfall more abundant, of starting a new Ice Age.

Suppose, on the other hand, that rising turbidity produced an offsetting worldwide drop in temperature. This, too, might start the glaciers crawling southward again.

Whichever way inadvertent weather modification goes, Bryson says, "it is inconceivable that the change would be beneficial."

A MISERABLE MESS

The evils of air pollution are more immediate than their long-term influence on climate. They damage human health, they kill livestock and crops. They make the earth an unlovely habitat for living creatures.

As Dr. Harrar of the Rockefeller Foundation said, "We have known for some time that man himself is the greatest single threat to his environment." As Dr. Blair of Texas University said, "Man has made a miserable mess of the world in which he lives."

What, by fouling the air, has man done to his own health? Photochemical smog was first defined in Los Angeles in the 1940s. The city, and the state of California, have struggled mightily ever since to do something about it.

Car makers have been forced to incorporate devices which lessen dangerous emissions. Controls have been placed on factories and refuse burners. And the upshot, according to the technical weekly *Science*, is that Los Angeles is about where it was a decade ago—new sources of pollution have kept pace with the effort to curb old ones.

In a single year, according to a staff report of the Senate Commerce Committee, Los Angeles doctors advised more than 10,000 patients to move somewhere else because of the harm being done to them by air pollution.

What is this harm? Clinically it is hard to pin down because even in the famous pollution "episodes," the victims had suffered previously from respiratory or heart maladies—the bad air just killed them sooner.

Nevertheless, according to the Senate report, air pollution "is a significant contributing factor" to the growing number of chronic diseases such as lung cancer, emphysema, bronchitis, and asthma.

City air is dirtier than country air, and statistics show that the death rate from lung cancer is 25 per cent higher in the cities than in rural regions.

A Purdue University study puts lung cancer deaths twice as high in large cities as in rural areas. It says polluted air has shortened city dwellers' lives, on a comparative basis, by five years.

In New York, New Jersey, California, Tennessee and elsewhere trees and crops downwind of the polluting sources have withered and died.

Growers of orchids and other flowers, of tobacco and garden produce, of citrus fruits and grapes have had to give up otherwise valuable farmlands because they were too close to air polluting cities. At least half the states have suffered crop damage from smog.

Air pollution, particularly sulfur oxides, has also been tough on materials. It corrodes and tarnishes metals, weakens and fades fabrics, makes leather and rubber brittle, discolors paint and stone, etches glass and messes up electrical circuits. It has damaged priceless old masters and great sculptures.

In all the air pollution disasters, dumb animals as well as sentient human beings have died. Says the National Tuberculosis and Respiratory Disease Association in a discussion of the Poza Rica, Mexico, hydrogen sulfide spill: "All the canaries and many other birds and animals died."

One of the pollutants particularly dangerous to animals is the fluorides released into the air from factories making fertilizer, aluminum, and iron and ceramics products.

Fluorides get into plants, which concentrate them. Cattle eat the plants. Their teeth become mottled. As they feed further "on this insidious food" they lose weight, give less milk, and finally become so crippled they have to be destroyed.

Is the United States doing anything about air pollution? It is. Following the lead of California, the federal government has issued auto emission control standards—not yet as stringent as California's—for the guidance of state governments.

Under the Air Quality Act of 1967 it is creating 57 control regions which by the summer of 1970 will include all the 50 states plus the District of Columbia, Puerto Rico, and the Virgin Islands. The states and other governing bodies are responsible for enforcing the clean air standards.

According to the Senate report, however, the proposed standards won't be effective. Said the committee:

"Studies indicate that, under existing controls, automobile air pollution in the United States will more than double in the next 30 years because of the projected increase in both the number of vehicles and miles driven by each vehicle."

One trouble is that health data aren't specific enough to prove air pollution is the villain doctors feel sure it is. You can't simulate in a laboratory the precise kind of polluted air most of us breathe.

Suppose it would cost $3 billion to $4 billion a year to curb air pollution in the next 10 years. Can you justify such expenditures for health reasons alone, forgetting the esthetic, crop, animal, material damage? Not at the moment, maybe.

But the health evidence is beginning to come in. In the Department of Health, Education and Welfare has been created the National Air Pollution Control Administration headed by Dr. John T. Middleton.

Commisioner Middleton forecast recently that accumulating data will lead eventually to far more stringent controls than any yet adopted.

So far there is little reason to believe that the air will be any cleaner 10 years hence than it is now, according to federal authorities. But attempts are being made to minimize the effects of bad air.

In Los Angeles, local broadcasting stations and newspapers for some time have included smog forecasts along with the daily weather report.

Now the U.S. Weather Bureau is starting special programs in St. Louis, Chicago, New York, Philadelphia, and Washington "in support of air pollution agencies." The special forecasts will warn of weather conditions which might aggravate pollution.

This will permit local officials to "take measures to control the amount of pollutants emitted from open dumps, smokestacks, and other pollution sources." The new Nimbus 3 weather satellite is equipped to measure pollutants as it looks down through the atmosphere from space.

Such measures may help to prevent disastrous "episodes." But it remains to be seen what effect they will have on the chronic smog which, according to the federal government, "is a serious threat to public health and welfare."

Not everybody agrees with those who say the automobile is a threat to mankind. According to auto industry spokesmen, emission control devices on new cars will, by the time used cars have left the highways, return the air to the relatively clean state of 1940.

The Senate report on auto emissions called for development of steam cars to reduce air pollution. Others have urged return of the smogless battery-driven cars of old.

Proponents of nuclear power contend they would cut down heavily on the amount of pollutants now dumped into the atmosphere by coal-fired power plants. Others are worried about radioactive contaminants that might be loosed if a nuclear plant failed.

If proliferating man cannot get all the energy he needs except by burning up his fossil legacies of coal and oil or by risking pollution by radioactivity, perhaps he should fall back on the original source, the sun.

Attempts to trap sunlight economically for power have not yet succeeded on any large scale. But Dr. Cole of Cornell feels the effort should continue.

Solar radiation, he has said, must become man's chief energy source—if the species lasts long enough to see this happen. In this connection, The National Geographic Society said recently that:

"If man could collect and efficiently use it, the sunlight falling on just the city of Los Angeles would supply more energy than is consumed in all the homes on earth."

It will be a long time before this comes to pass. Meanwhile, where are we? The National Tuberculosis and Respiratory Disease Association in its "pollution primer" says:

"Nature is fighting a losing battle with man-made air pollution . . . vast expanses of countryside smolder and stink. Dreamy fogs are accomplices to murder. Sunny, windless days carry, like a disease, the threat of suffocation."

Why should we care? The Environmental Pollution Panel of the President's Science Advisory Committee said this:

"Man is but one species living in a world with numerous others; he depends on many of these others not only for his comfort and enjoyment but for his life."

Says climatologist Reid A. Bryson of Wisconsin:

"We would like our grandchildren to experience blue skies more often. . . ."

The Tuberculosis Association summed it up:

"Air pollution threatens not only man's wallet and his health. Air pollution erodes his soul. Every mountain blacked out by pollution, every flower withered by smog, every sweet-smelling countryside poisoned by foul odors destroys a bit of man's union with Nature and leaves his spirit diminished by loss."

Perhaps the simplest summation of all was uttered by Russell E. Train, president of the Conservation Foundation, who became Undersecretary of Interior in the Nixon Administration.

"The real stake," he said, "is man's own survival—in a world worth living in."

NOISE

FRUITS OF PROGRESS CAN BE BITTER AS WELL AS SWEET

The racket may not be killing you. On the other hand, maybe it is—slowly, insidiously.

The din that assaults our ears almost non-stop is shattering our tranquility, hurting our health, and contaminating an environment already poisoned by air, water and soil pollution.

Pollution of our increasingly despoiled living space by noise is another example of the now widely recognized truth that the fruits of technology can be bitter as well as sweet.

Noise pollution, which has been called "the price of progress," is getting worse every year. Nothing very effective is being done about it—in this country, at least. According to Sen. Mark O. Hatfield, R-Ore., the United States is the noisiest of modern societies.

The consensus at an American Medical Association Congress in Chicago was that noise is as great a hazard to mankind as air and water pollution. It does both physical and psychological damage.

One of the speakers at the AMA Congress said: "The public must be made aware that offensive noise can be controlled and must be made angry enough to do something about it."

Noise has been defined in many ways. It is unwanted sound, sound without value, unrestricted sound, sound that hurts, harms, distracts, destroys sleep, invades privacy, frightens, irritates, or simply annoys.

Just how dangerous is the clatter and clamor and the mechanical screeching and screaming which assail us all and from which there appears to be no escape, not even in suburbia?

Dr. Vern O. Knudsen, Chancellor Emeritus of the University of California and a distinguished student of acoustics, the science of sound, has given this answer:

"Noise, like smog, is a slow agent of death. If it continues for the next 30 years as it has for the past 30, it could become lethal."

Whether our environmental noisiness actually can kill us is debatable. But the Federal Council for Science and Technology, a White House agency, notes that "growing numbers of researchers fear that the dangerous and hazardous effects of intense noise on human health are seriously underestimated."

There is no doubt that industrial din has inflicted loss of hearing on millions of workers. At least a million workers now living suffer from some degree of deafness. The Federal Council estimates that another 6 to 16 million are exposed to noise levels which may ruin their hearing in the future.

Deafness, in fact, has finally been recognized as an occupational hazard in a lot of major industries.

But more alarming than industrial racket, because millions more persons are affected, is the steeply rising level of "community noise" which afflicts everybody—in homes, offices, schools, hospitals, even vacation resorts.

Some authorities believe that noise, along with crowding, has triggered calamities that might not have happened without the spur of noise.

Ailments which may have been caused or at least aggravated by noise include ulcers, heart diseases, allergies, and mental illness. Foreign reports have even attributed sexual impotency to high noise levels in factories.

Racket can be dangerous in indirect ways. For example: when it drowns out alarm signals or shouted warnings or verbal instructions vital to safety.

It can distract attention and interfere with the vigilance of persons responsible for monitoring controls and reacting instantly to danger signs.

A startling noise, such as a sonic boom, conceivably could cause a surgeon's knife to slip.

It has seriously been proposed that noise, as a legacy of the presumably uncaring technological "Establishment," has been implicated in some way in ghetto and campus rioting.

Paul N. Borsky of the Columbia University School of Public Health says that if a person feels the noise makers are concerned about his welfare and are trying to muffle the din, he is likely to remain tolerant.

"If he feels, however, that the noise propagators are callously ignoring his needs and concerns, he is more likely to be hostile . . . .

"This feeling of alienation, of being ignored and abused, is also the root cause of many other human annoyance reactions.

"This is one of the major reasons cited for urban riots, discontent by minority groups, and more recently of student revolts."

A similar idea has been expressed by Joseph J. Soporowski, Jr., environmental scientist of Rutgers College. He says:

"Each of us can perhaps recall responding with indignation to assaults upon our freedoms. Yet, until recently, we have failed to respond similarly to equal assaults upon the delicate mechanism of our ears. Perhaps the cause of some of our problems and differences could be traced to irritating noise."

Warning of the "dangerous din to come," Soporowski said "little is being done to curb this potential menace; little is being done to halt pollution of our environment by noise."

U.S. LAGS FAR BEHIND

"The overall loudness of environmental noise is doubling every 10 years in pace with our social and industrial progress."

This is the conclusion of the Federal Council for Science and Technology, one of the many groups looking into damaging effects of noise in our modern society.

"Immediate and serious attention must be given to the control of this mushrooming problem," says the Council, because otherwise "the cost of alleviating it in future years will be insurmountable."

What about the charge that "little is being done to curb this potential menace"?

"There is no doubt," says the Council, "that recognition of the noise problem in America has arrived late. With the exception of aircraft noise, the United States is far behind many countries in noise prevention and control."

It took the Donora smog disaster to awaken Americans to the horrors of air pollution. In October, 1948, a poisonous pall enveloped Donora, a town of 14,000 in the heavily industrialized Monongahela River valley of Pennsylvania. Some 6,000 Donorans were sickened and 20 died.

Will it take another Donora to alert us to the dangers of noise pollution?

The nation's first National Conference on Air Pollution was held in Washington in 1958—10 years after the Donora tragedy. In June a year ago the first National Conference on "noise as a public health hazard" was held, also in Washington. U.S. Surgeon General William

H. Stewart noted that "we haven't had our Donora episode in the noise field."

"Perhaps," Stewart continued, "we never will. More likely, our Donora incidents are occurring day by day, in communities across the nation—not in terms of 20 deaths specifically attributable to a surfeit of noise, but in terms of many more than 20 ulcers, cardiovascular problems, psychoses, and neuroses for which the noises of 20th century living are a major contributory cause."

Much remains to be nailed down about the multitude of ways in which noise hurts our health and efficiency and serenity.

But, asks the Surgeon General, "Must we wait until we prove every link in the chain of causation?"

"In protecting health," Stewart said, answering his own question, "absolute proof comes late. To wait for it is to invite disaster or to prolong suffering unnecessarily."

As noted earlier, noise has been called "the price of progress," the technological progress that has given us so many of the things we value, from air conditioning to vacuum cleaners. It also has given us some things we loathe, such as the cheaply built apartment houses which sound specialist Leo L. Beranek of the Massachusetts Institute of Technology calls "acoustical torture chambers."

According to the Federal Council for Science and Technology, old-fashioned dwellings of 40 to 50 years ago "were comparatively quiet places in which to live." Thanks to modern construction techniques we have "some of the noisiest buildings in existence."

Rep. Theodore R. Kupferman, R-N.Y., says New York City is busily building "the noise slums of the future."

Noise, of course, is not new. In the London of 1800 the quacking of fowl and the bellowing of animals being driven through the streets to slaughter made life hideous for people already fed up with the ceaseless clip-clop of horses' hooves on cobblestones.

Now, thanks again to technological progress, we have unmuffled scooters, motorbikes, sports cars, trucks; jet aircraft, sonic booms, screeching tires, sirens, jackhammers, air compressors; a host of kitchen and other household contrivances that whine, clank, gurgle, or rattle.

We have also clattering typewriters and cackling secretaries, the neighbors' radio and television sets, the startling crescendo of suddenly turning-on building air conditioners; the endless ringing of telephones, the din of the pile drivers, bulldozers, power saws, lawn mowers; the conversation-killing sound of dull music piped into elevators or restaurants, the distracting bells of ice cream wagons, the giant insect hum of the upstairs tenant's vacuum, the flushing of other people's toilets and the drawing of other people's baths.

Also painfully familiar to all are the night noises that sound like pistol shots but may only be backfires, puzzling and sleep destroying; the round-the-clock squawking of auto horns, the keening of handheld transistor radios; the bone-tingling throb of the electric guitar, the cacophony of rock and roll bands, and the dawn chorus of the garbage collectors with their special brand of crash-bang basketball.

Make your own list. To do damage of one kind or another, a sound doesn't have to be loud—it just has to be unwanted. A Washington, D.C., psychiatrist once remarked that he was going to get a shotgun and slay the mocking bird that kept him awake with its tweedling in the night hours.

He may have been joking, or perhaps he needed a psychiatrist himself. But the serious fact remains that all of us are captive audiences: we are forced to listen to sounds—and not just mockingbird song—that we can't escape.

A favorite cliché among scientists of sound is that "one man's music may be another man's noise." Church chimes, if what you need is silence, can torture.

There are those who feel sure they love and must have hi-fi and rock and roll music, which generates noise far above the levels that would be forbidden in a well-run boiler shop.

A number of acoustics experts, alarmed by increasing evidence of hearing loss among the young, have suggested it might be a good public health measure to make discotheques display entrance signs reading:

"Caution: the noise levels inside may be hazardous to your health."

NOISE CAN BE MUFFLED—FOR A PRICE

Hearing loss caused by noise, according to the Federal Council for Science and Technology, constitutes "a major health hazard in American industry."

Industries guilty or suspected of inflicting hearing hazards on their workers, according to the Council, "include iron and steel making, motor vehicle production, textile manufacturing, paper making, metal products fabrication, printing and publishing, heavy construction, lumbering and wood products, and mechanized farming."

To this list may be added such military functions as "flight line and carrier deck operations, engine test cells and weapons firing, armor operations, and assorted repair and maintenance work."

No general, universally applicable standard of safe noise exposure exists in the United States. But Secretary of Labor George P. Shultz in May promulgated regulations under the Walsh-Healey Act for work done on federal contracts over $10,000 in value.

The regulations set an allowable level of industrial noise at 90 decibels for prolonged exposure (eight hours a day).

The decibel is a unit of sound used by scientists. The smallest sound an acute human ear can perceive is about one decibel. Ordinary breathing registers 10 on the decibel scale. Rustling leaves shoot the count up to 20. A restaurant where the decibel level is only 50 is quiet indeed. Other decibel values:

Conversation 60, heavy traffic (or an office with tabulating machines) 80, food blender 93, an amplified rock and roll band 138, the pain level for human ears 140, jet plane takeoff 140, space rocket liftoff 175.

Since the decibel scale is logarithmic, a noise high in the list (rocket) may be billions of times more powerful than one near the bottom (breathing).

Authorities appear to agree that most sounds of less than 40 decibels (the normal sustained noise inside a residence when the hi-fi and the blenders and disposal systems aren't going) are hardly noticeable.

But long exposure to levels above 80 decibels can damage the sensitive apparatus of the ear.

According to the Federal Council, "traffic noise radiating from the freeways and expressways and from midtown shopping and apartment districts of our large cities probably disturbs more people than any other source of outdoor noise." (Aircraft noise is more intense but exposure time is less than round-the-clock highway noise.)

And of all vehicles, "the trailer truck is perhaps the most notorious noise producer." At expressway speeds, a single truck may generate noise above 90 decibels while a long line of truck traffic may produce levels above 100.

At a noise conference in Washington a year ago [1968] it was noted that many trucks come equipped with fairly adequate mufflers. But operators often destroy the mufflers under the impression that they reduce efficiency.

It costs money to suppress noise. The aircraft industry is spending millions in this attempt.

But not suppressing noise also costs money. Sen. Mark O. Hatfield, R-Ore., reported recently that the cost of noise to industry generally— in compensation, lost production, decreased efficiency—is estimated "at well over $4 billion (repeat billion) per year."

This loss is accumulated in bits and pieces and in subtle ways. The experts say 19 per cent more energy is needed to do a job in a noisy place than in a quiet one.

Everybody says noise can be muffled—if we are willing to pay the price for doing it. Other countries, but not the United States, have included national sound insulation regulations in their building codes.

These nations are Austria, Belgium, Bulgaria, Canada, Czechoslovakia, Denmark, England, Finland, France, Germany, the Netherlands, Norway, Scotland, Sweden, and the U.S.S.R.

"In the field of architectural acoustics and the control of noise in buildings," says the Federal Council, "we are far behind federally supported or implemented research in Canada, England, and Europe, and are currently behind the level of activity of Russian and Japanese theoretical, analytical, and applied research."

London, Berlin, and Paris have decreed leather or rubber rims on their garbage cans to curb one of the greatest noise annoyances of city life.

But only recently has New York begun even to experiment with plastic or paper bags as a means of eliminating the jangling danger of the traditional metal cans.

Some states are trying to cut down on highway cacophony. Connecticut plans to set up "sound traps" similar to the radar "speed traps" now in use.

Nonetheless, noise is escalating, along with the number of physical and mental maladies attributed to it.

Coming up is the supersonic transport which, if it ever is put into transcontinental service, will smite the eardrums of up to 50 million Americans daily with the clap of doom otherwise known as the sonic boom.

There are those who feel that only city and state regulations can deal effectively with the rising tide of din. But the federal government, say others, has the responsibility for doing noise research and providing noise "guidelines" for local governments.

What is needed, says the Federal Council, is "a total national program to abate undesirable noise."

As W. H. Ferry, vice president of the Center for the Study of Democratic Institutions, Santa Barbara, has noted, noise like the sonic boom "is de-civilizing."

"No self-respecting civilization," says Ferry, "ought to have to accommodate itself to such an annoyance."

UGLIFICATION

We have been warned: our garbage, our junk, our rubble threaten to engulf us.

We have devoted much thought to pollution of the air we breathe and to the water we drink.

But we have paid little attention, comparatively, to what the nice-minded call "solid wastes."

Yet solid wastes in their myriad forms—everything from animal dung heaps and city garbage to universal litter and abandoned autos—are the worst of the polluters.

They pollute not only air and water but the landscape. They add "uglification" to the mess of horrors man has contrived for himself. They also constitute a reckless waste of irreplaceable resources.

If not a tribute, they are at least a monument to our affluence, our technological ingenuity, and our "heritage of waste" in a use-and-discard society.

As technology presents us with ever more conveniently packaged "consumer items" and as man's numbers mushroom, the rubbish pile grows ever higher. It is, in fact, growing faster than the population.

Charles C. Johnson Jr., administrator of the new Consumer Protection and Environmental Health Service of the Public Health Service, states it this way:

"Growing mountains of garbage and trash threaten to bury us in our own waste products."

They already are hurting our health. They already have destroyed large areas of living space which nature had allotted to creatures of the wild. They already have spread "scenic blight" throughout the countryside. They have contributed their large bit to what Johnson says is a rapidly approaching drinking water crisis.

It used to be when the nation was young that no harm was done if you just threw away something you no longer wanted, if you just "spread around" your garbage.

But as the National Academy of Sciences has pointed out in a special report, "As the earth becomes more crowded, there is no longer an 'away.' One person's trash basket is another's living space."

According to Johnson, this country is now trying to deal with 3.5 billion tons of solid wastes every year. This includes 1.5 billion tons of animal excreta, 550 million tons of what's left over from the marketable parts of farm crops, 1.1 billion tons of mineral wastes, 100 million tons of industrial trash, and 250 million tons of household, commercial, and municipal wastes.

These figures do not include the millions of automobiles junked each year. It has been estimated that the car discard rate will reach eight million a year by 1975.

In addition to all this is the unguessable (certainly in the billions of tons) of annually accumulated debris from the demolition of buildings and highways to make way for new ones.

Each of us contributes on the average 5.3 pounds to the garbage man's haul of food scraps or rubbish. Every 30 seconds some one of us abandons a dead automobile on a city street or country roadside.

Abandoned cars are a familiar part of big city litter. About 120 a day are left on New York streets for the city to haul away.

There are now between 10 million and 30 million junked cars lying about the country, disfiguring the landscape or congesting automobile graveyards.

The environs of main highways across the nation harbor some 17,500 junkyards populated largely by auto hulks.

Just to get rid of household, municipal, and industrial refuse costs us about $4.5 billion a year. Of all municipal costs, this is exceeded only by what we pay for schools and roads.

But 85 per cent of this annual expenditure goes solely for collection, with only about 15 per cent spent for ultimate disposal. According to one estimate, the United States would have to spend another $3.75 billion in the next five years to provide a suitable system of waste disposal.

As things stand, according to Charles Johnson of the Environmental Health Service, we "have not yet figured out what to do with the refuse that litters our countryside."

More than half of the nation's communities over 5,000 in population dispose of their wastes in a fashion described by the Public Health Service as "improper." Open dumping accounts for nearly 80 per cent of all waste disposed of in this country. The Academy of Sciences report deplored this practice.

"Too often, refuse-disposal areas are open dumps—festering and disfiguring the landscape," the report said, "Flies, rats, and other disease-carrying pests find large quantities of food and suitable harborage in the piles of exposed refuse.

The polluted drainage from open dumps is an additional insult to adjacent ground and surface water supplies. Characteristic foul odors, produced by decomposition, together with the smoke created by inefficient open burning, are often identifiable for miles."

Every cubic foot of garbage, it has been estimated, produces about 75,000 flies, not to mention rats, mice, mosquitoes, cockroaches, and other unlovely pests.

The great cities with their incinerators and "sanitary landfills" have progressed a little beyond the open dump disposal system. But the general "state of the art" remains about what it was 50 years ago. It has been said that the last real invention in waste disposal was the garbage can, and that the most recent improvement was putting an engine instead of a horse in front of the garbage truck.

Disposal means different things to different people. A housewife in a city apartment disposes of garbage effectively enough for her needs by shoving it down the incinerator shaft, or dumping it into a garbage can out front or in the alley.

The waste disposal chore is greatest in the cities where 70 per cent of the national population dwells on 10 per cent of the land. And even in the cities, the enormity of the problem is brought home to the householder only when the garbage men strike.

A few days without garbage collection make the streets unnavigable by the fastidious. Many of the nation's smaller communities have no regular collection services.

Considering the health hazards involved, it may be a wonder even the larger ones do. According to Richard D. Vaughan of the Environmental Control Administration, garbage collectors "are engaging in one of the most dangerous occupations in existence."

Experts agree there is only one ultimate solution to the solid waste disposal dilemma. They call it "total recycling."

This is a dream of a nearly junkless society. In it, nothing would ever be thrown away; it would be used again. Our wastes then would become a national resource, a "mother lode" of valuable materials.

Automobiles, for example, would be designed either for reuse or for easy retrieval of their better parts. When automobiles had served out their useful lifetimes, the assembly-line process that produced them would be reversed.

Run backwards through the line, they would yield their most precious parts in a sort of priority system until only irreducible scrap, itself salvageable, remained.

But this is not the way auto makers design autos. Nor is it the way mercantile companies design the packages with which they lure the consuming public.

The modern steel industry no longer has to have scrap iron. And the people who package foods and everything else for the American family have no economic reason for caring what happens to the empty package.

City trash is a fantastic mixture. The most revolting part—garbage —is the easiest to get rid of. If man simply ignored it, taking the consequences to his eyes, nose, and health, nature would dispose of it.

But some of the stuff mixed with garbage in the trash haul is what scientists call "non-degradable." You can bury a nylon stocking in moist soil for years, and when you dig it up, there it is. Soil bacteria and other organisms which feast on garbage can't stomach such synthetic materials as nylon and plastics generally.

In a typical year Americans throw away 48 billion cans, 26 billion bottles, more than 30 million tons of paper, four million tons of plastics, and 100 million wornout tires weighing a million tons.

To simplify life for Americans caught away from home without an opener, technology provided the "snap top" beer and soft drink can. It was made of aluminum because aluminum cans are easier to make and tear open than cans of steel.

But aluminum is more resistant to corrosion than steel and hence harder for nature to reduce to rust. Another bit of commercial progress is the non-returnable glass bottle. Instead of lugging it back to the store for the deposit, you pitch it into the trash can along with the potato peelings.

The non-returnable bottle is looked upon almost as an enemy by those who preach "reuse and recycling" as the best answer to waste disposal problems. The committee on pollution of the National Academy of Sciences posed this question:

"Should we tax glass bottles severely, or have federal law 'forbid' that they be not reused?"

According to Solid Wastes Management magazine, it costs New York State 30 cents for each bottle it picks up. This is seven times what it costs to make the bottle in the first place.

Such inorganic wastes in the trash mountains complicate the task of disposal. Plastics, brick, and concrete, the Academy of Sciences committee noted, "may endure for centuries."

The aluminum can, the throw-away bottle, and the plastic container have contributed more than they were ever worth to "landscape pollution." On a Sunday afternoon the Washington Cathedral grounds on Mt. St. Albans in Washington, D.C., is cluttered with cans.

City dwellers walking to the bus stop have to tread their way gingerly amongst the shards of glass bottles which have been flung to the sidewalks from automobiles.

There is no escaping the litter—or the conclusion that Americans are incorrigible litterers. On some highways it is hard to see the "no dumping" signs protruding from the cascades of dumped refuse.

According to *The New York Times*, school kids who went fishing in New York's Central Park pond for wildlife caught discarded auto tires, glass containers, beer cans, waterlogged magazines, and a blanket.

Prominently placed trash cans bearing legends reading "keep your city clean" have helped, but not enough. Where man goes, he leaves litter.

Those who have toiled gasping to the summit of Colorado's 14,256-foot Longs Peak have found awaiting them a refuse can. Some hardy ranger lugged it there.

His work was not altogether in vain; on an ordinary summer's day there is, indeed, much trash in the can. But there is litter elsewhere—remnants of sandwiches, candy bar wrappers, drained milk containers just lying around or wedged amongst the rocks in this space so high above and so far removed from the normal range of the litterbug.

According to the Senate subcommittee on air and water pollution, "perhaps the greatest waste collection headache presented by packaged materials is littering along our roadways, in our parks, and along our rivers and lakes.

"There is a vast difference in costs," says the subcommittee, "between collecting a ton of cigarette wrappers placed in garbage cans and a ton thrown away carelessly."

Authorities now agree that disposal of solid wastes must be accomplished on a regional urban-suburban-rural basis. The old local community attitude of "take it somewhere else, but don't raise my taxes in the process" no longer is tolerable, according to a National Academy of Sciences study.

And as Charles C. Johnson Jr., administrator of the Consumer Protection and Environmental Health Service, notes: "Yesterday's city dump is now in today's suburb."

The traditional methods of disposing of big city wastes are incineration and landfill. But most city incinerators are inefficient burners, and they compound the sin of air pollution.

At best, they just reduce the volume of waste. Anywhere from five to 25 per cent of solid wastes, depending on incineration efficiency, remain to be stowed some other way.

The "sanitary landfill" has had some notable successes—and some notable failures. The landfill has been used to create parks, recreational areas, and public gardens. Ideally, you pick a site which has no other use, bulldoze trenches in it, haul in compacted or shredded refuse, and cover the unsightly stuff daily with decent soil.

In practice this often has resulted in seepage of pollutant matter into ground water supplies, or in destroying the marsh habitats of wild things already losing too many struggles against encroachment by man.

As Johnson says, "most cities in the country are now destroying out-of-the-way areas of natural beauty, and polluting land, air, and water in an effort to get rid of mountains of refuse."

But landfill, though a solution of diminishing usefulness as sites become scarcer, has accomplished some fine things. In Los Angeles County an open pit mine was filled with wastes to provide a botanical garden of great beauty.

In Detroit solid wastes have been fashioned into an artificial mountain for sledding and skiing in winter. Virginia Beach, Va., is converting an old landfill site into a soap box derby slope and outdoor theater.

There are other examples, but they cannot be endlessly repeated in the long future. Space is running out. It is becoming more and more necessary to haul city trash to increasingly distant disposal sites, on the land or in the sea.

San Francisco is studying a plan to haul garbage 300 miles by rail to desert burial grounds instead of dumping it as before into the bay. Philadelphia is putting into operation a program for transporting its refuse to abandoned mines 100 miles away.

For Philadelphia this will be cheaper by $2 a ton than incineration. There are some 8,000 abandoned mines in this country. Sooner or later they may all be filled with the junk of civilization.

There remains the sea. It has been suggested that offshore islands might be built of city wastes in the Atlantic for use as supersonic aircraft runways.

Frank R. Bowerman of Zurn Industries, writing in the *Investment Dealers' Digest*, has estimated that one runway could be built every year in shallow offshore waters with the eight million tons of solid wastes produced annually by New York City.

The University of Rhode Island is examining the possibility of burning city garbage in incinerator ships which would dump the ashes into the sea. This would solve the problem of the old garbage scow, the fruits of which too often washed back ashore, polluting the beaches.

New York scientists meanwhile made a discovery which may prove significant in the future—garbage tossed into the offshore waters attracted hordes of fish which, unhappily for fishermen, have moved elsewhere since this practice was abandoned.

The New York sanitation department dumped 500 junked automobiles in the ocean off Long Island last year to find out whether auto hulks would serve as artificial habitats and breeding grounds for fish. This idea has been recommended by students of the dead auto problem and even was promoted in a comic strip recently.

Who knows? Maybe it will save New York's offshore fisheries and at the same time provide an answer to what to do with old autos that have not yet found a suitable burying ground. Deep ocean canyons also have been recommended as storage for things no longer wanted on land.

A lot of other ideas have been suggested and many of them tried. Europe, perhaps because it is smaller and more crowded, has tried harder and gone further than the United States toward perfecting efficient incinerators.

The Japanese have invented a new "dense compaction" process which presses mounds of ordinary refuse into blocks of a ton or more that will sink in water. At one time it was hoped such blocks could be used as building materials.

But gases generated by the organic material within appear to have made this an impractical and even dangerous technique. Dense compaction does, however, promise to be a boon for long-haul cleanup trains bearing trash from cities to distant disposal sites.

Much effort has been made to enlist the profit motive in the battle to save the environment from man's wastes. Tried many times but still found wanting is "composting," the decomposition of city and farm wastes through bacterial action for production of fertilizer and humus for agriculture.

The competition of more convenient chemical fertilizers has all but killed composting for profit. This also explains why the great cattle feedlots of the nation are generating piles of animal wastes faster than they can get rid of them.

Time was when organic material of this sort, spread over the fields, kept our soil rich and productive. Now there is little to do with it except let its odors befoul the air and its drainage pollute our waters.

About two-thirds of U.S. beef production comes from cattle feedlots. One cow produces waste equal to the sewage of 16 people. One feedlot handling 10,000 head of cattle has the same waste disposal needs of a city of 160,000 persons.

According to former Secretary of Agriculture Orville L. Freeman, the nearly three million head of cattle fed on lots in Nebraska and Iowa alone create an amount of waste equal to that produced by 49 million human beings, 11 times the population of the two states.

Farmers who once drove a manure spreader back and forth across their fields now find it cheaper and easier to scoop the fertilizer they need out of a bag than to dig it out of a dung pile.

Authorities are becoming convinced that salvaging human, industrial, animal, commercial and other wastes should be considered a means of cutting disposal costs rather than a way to make a profit.

Hempstead, N.Y., salvages heat from its refuse incinerator to run a 2,500-kilowatt electric power station and a 420,000 gallon a day desalting plant. But the net effect is to make waste disposal less costly, by no means to make it profitable.

Salvage probably is one of man's oldest occupations. Many persons now living remember that every little town once was roughly divided by the railroad tracks.

On one side were the stores and the homes and the schools and churches, on the other the junkyards where a reasonably active lad could collect a dime a week or more by turning in scavenged scraps of anything from tinfoil to babbitt metal or discarded copper wire.

But the small town junkyards are disappearing. In their place is a $3 billion a year industry of at least 2,300 large companies. This industry, however, is interested not in municipal wastes but in clean and easily identifiable commercial and industrial discards.

Residential trash is too mixed to be worth the modern junkman's attention. The most commonly salvaged solid wastes now are such things as cardboard, newspapers, steel cans, auto bodies, and a small amount of glass bottles.

Broken glass occasionally is salvaged and crushed for remelting into new bottles and jars. Some research is being done into use of smashed glass as a substitute for sand or gravel. The possibility has been raised of inventing a kind of bottle that will automatically dissolve after it has been emptied.

There are those who believe that solid waste disposal is the most serious of the pollution problems man has brought upon himself. This is partly because it creates both air and water as well as land pollution, and partly because mankind obviously is losing the race against burial under its own piling up discards.

If you don't believe this, look around you and try to think what the scene will be like 20 or 40 or 100 years hence, with the population doubled or tripled, or quadrupled, if some new solutions aren't conjured up in the meantime.

The government is trying to forestall disaster with a program for financing research into waste disposal schemes and assisting in local and regional plans for building demonstration plants incorporating modern disposal methods.

According to Charles C. Johnson Jr., director of the Consumer Protection and Environmental Health Service, the solid waste "environmental problem may well prove the most difficult and serious of all."

One difficulty is that waste disposal has been considered a local responsibility and, at the same time, a local irresponsibility. Communities adjacent to cities, the big waste producers, have tended to shrug off their problems as being beyond their own control.

So the federal government is attempting to distribute its research and demonstration grants as far as possible on a regional or interstate basis while at the same time not refusing assistance to communities with special local difficulties.

The goal, if man is not finally to be swamped by his own filth, is to get new use out of everything he has ever used. This is the gospel of "recycle and reuse." It is more than just an antidote to suffocation in garbage and litter.

It is the only ultimate answer to the fact that man is throwing away priceless resources he cannot recapture except by using them over and over again.

It is reminiscent of the old New England saying: "Use it up, wear it out, make it do."

This is the "economic" approach. There is another, the humane. Says Johnson of the Environmental Health Service:

"In the inner city, accumulated garbage and trash create breeding grounds for rats, insects, and vermin, and constitute a major health problem.

"Before we can do anything effective in the deteriorating areas of our cities," Johnson says, "we have to attack the problem of solid waste disposal."

There is yet another, more universal, way of looking at what all of us in our profligate carelessness are doing to ourselves and to those who will inherit our soiled world. Again Johnson:

"We must halt the deterioration of the environment. We must make life worth living in the ghetto and in the suburbs, in the town house and in the cottage, in the city and in the country.

"We must prove that ugliness, danger, and misery do not have to be a part of the birthright of any American, wherever he may live in this land."

NATURE
FIGHTS BACK *

RACHEL L. CARSON

7

Rachel Carson, author of "The Sea Around Us," "The Edge of
the Sea," and "Under the Sea-Wind," was one of America's
foremost biologists. In all her work, Miss Carson's basic interest was the relation of life to its environment. From 1958 until
her death a short time ago, she collected data from scientists
all over the world about insecticides and their effects on the
living community. The result was "Silent Spring," probably
the most important book on ecology of the century. The book
is an eloquent protest in behalf of the unity of all nature, a
protest in behalf of life.

To have risked so much in our efforts to mold nature to our satisfaction and yet to have failed in achieving our goal would indeed be the final irony. Yet this, it seems, is our situation. The truth, seldom mentioned but there for anyone to see, is that nature is not so easily molded and that the insects are finding ways to circumvent our chemical attacks on them.

"The insect world is nature's most astonishing phenomenon," said the Dutch biologist C. J. Briejèr. "Nothing is impossible to it; the most improbable things commonly occur there. One who penetrates deeply into its mysteries is continually breathless with wonder. He knows that anything can happen, and that the completely impossible often does.'

The "impossible" is now happening on two broad fronts. By a process of genetic selection, the insects are developing strains resistant to chemicals. . . . But the broader problem, which we shall look at now, is the fact that our chemical attack is weakening the defenses inherent in the environment itself, defenses designed to keep the various species in check. Each time we breach these defenses a horde of insects pours through.

From all over the world come reports that make it clear we are in a serious predicament. At the end of a decade or more of intensive chemical control, entomologists were finding that problems they had considered solved a few years earlier had returned to plague them. And new problems had arisen as insects once present only in insignificant numbers had increased to the status of serious pests. By their very nature chemical controls are self-defeating, for they have been devised and applied without taking into account the complex biological systems against which they have been blindly hurled. The chemicals may have been pretested against a few individual species, but not against living communities.

In some quarters nowadays it is fashionable to dismiss the balance of nature as a state of affairs that prevailed in an earlier, simpler world —a state that has now been so thoroughly upset that we might as well forget it. Some find this a convenient assumption, but as a chart for a course of action it is highly dangerous. The balance of nature is not the same today as in Pleistocene times, but it is still there: a complex, precise, and highly integrated system of relationships between living things which cannot safely be ignored any more than the law of gravity can be defied with impunity by a man perched on the edge of a cliff. The balance of nature is not a *status quo*; it is fluid, ever shifting, in a constant state of adjustment. Man, too, is part of this balance. Sometimes the balance is in his favor; sometimes—and all to often through his own activities—it is shifted to his disadvantage.

Two critically important facts have been overlooked in designing the modern insect control programs. The first is that the really effective control of insects is that applied by nature, not by man. Populations are kept in check by something the ecologists call the resistance of the environment, and this has been so since the first life was created. The amount of food available, conditions of weather and climate, the presence of competing or predatory species, all are critically important. "The greatest single factor in preventing insects from overwhelming the rest of the world is the internecine warfare which they carry out among themselves," said the entomologist Robert Metcalf. Yet most of the chemicals now used kill all insects, our friends and enemies alike.

The second neglected fact is the truly explosive power of a species to reproduce once the resistance of the environment has been weakened. The fecundity of many forms of life is almost beyond our power to imagine, though now and then we have suggestive glimpses. I remember from student days the miracle that could be wrought in a jar containing a simple mixture of hay and water merely by adding to it a few drops of material from a mature culture of protozoa. Within a few days the jar would contain a whole galaxy of whirling, darting life—uncountable trillions of the slipper animalcule, *Paramecium*, each small as a dust grain, all multiplying without restraint in their temporary Eden of favorable temperatures, abundant food, absence of enemies. Or I think of shore rocks white with barnacles as far as the eye can see, or of the spectacle of passing through an immense school of jellyfish, mile after mile, with seemingly no end to the pulsing, ghostly forms scarcely more substantial than the water itself.

We see the miracle of nature's control at work when the cod move through winter seas to their spawning grounds, where each female deposits several millions of eggs. The sea does not become a solid mass of cod as it would surely do if all the progeny of all the cod were to survive. The checks that exist in nature are such that out of the millions of young produced by each pair only enough, on the average, survive to adulthood to replace the parent fish.

Biologists used to entertain themselves by speculating as to what would happen if, through some unthinkable catastrophe, the natural restraints were thrown off and all the progeny of a single individual survived. Thus Thomas Huxley a century ago calculated that a single female aphis (which has the curious power of reproducing without mating) could produce progeny in a single year's time whose total weight would equal that of the inhabitants of the Chinese empire of his day.

Fortunately for us such an extreme situation is only theoretical, but the dire results of upsetting nature's own arrangements are well known to students of animal populations. The stockman's zeal for eliminating the coyote has resulted in plagues of field mice, which the coyote formerly controlled. The oft repeated story of the Kaibab deer in Arizona is another case in point. At one time the deer population was in equilibrium with its environment. A number of predators—wolves, pumas, and coyotes—prevented the deer from outrunning their food supply. Then a campaign was begun to "conserve" the deer by killing off their enemies. Once the predators were gone, the deer increased prodigiously and soon there was not enough food for them. The browse line on the trees went higher and higher as they sought food, and in time many more deer were dying of starvation than had formerly been killed by predators. The whole environment, moreover, was damaged by their desperate efforts to find food.

The predatory insects of field and forests play the same role as the wolves and coyotes of the Kaibab. Kill them off and the population of the prey insect surges upward.

No one knows how many species of insects inhabit the earth because so many are yet to be identified. But more than 700,000 have already been described. This means that in terms of the number of species, 70 to 80 per cent of the earth's creatures are insects. The vast majority of these insects are held in check by natural forces, without any intervention by man. If this were not so, it is doubtful that any conceivable volume of chemicals—or any other methods—could possibly keep down their populations.

The trouble is that we are seldom aware of the protection afforded by natural enemies until it fails. Most of us walk unseeing through the world, unaware alike of its beauties, its wonders, and the strange and sometimes terrible intensity of the lives that are being lived about us. So it is that the activities of the insect predators and parasites are known to few. Perhaps we may have noticed an oddly shaped insect of ferocious mien on a bush in the garden and been dimly aware that the praying mantis lives at the expense of other insects. But we see with understanding eye only if we have walked in the garden at night and here and there with a flashlight have glimpsed the mantis stealthily creeping upon her prey. Then we sense something of the drama of the hunter and the hunted. Then we begin to feel something of that relentlessly pressing force by which nature controls her own.

The predators—insects that kill and consume other insects—are of many kinds. Some are quick and with the speed of swallows snatch their prey from the air. Others plod methodically along a stem, plucking off and devouring sedentary insects like the aphids. The yellowjackets capture soft-bodied insects and feed the juices to their young. Muddauber wasps build columned nests of mud under the eaves of houses and stock them with insects on which their young will feed. The horseguard wasp hovers above herds of grazing cattle, destroying the blood-sucking flies that torment them. The loudly buzzing syrphid fly, often mistaken for a bee, lays its eggs on leaves of aphis-infested plants; the hatching larvae then consume immense numbers of aphids. Ladybugs or lady beetles are among the most effective destroyers of aphids, scale insects, and other plant-eating insects. Literally hundreds of aphids are consumed by a single ladybug to stoke the little fires of energy which she requires to produce even a single batch of eggs.

Even more extraordinary in their habits are the parasitic insects. These do not kill their hosts outright. Instead, by a variety of adaptations they utilize their victims for the nurture of their own young. They may deposit their eggs within the larvae or eggs of their prey, so that their own developing young may find food by consuming the host. Some attach their eggs to a caterpillar by means of a sticky solution; on hatching, the larval parasite bores through the skin of the host. Others, led by an instinct that simulates foresight, merely lay their eggs on a leaf so that a browsing caterpillar will eat them inadvertently.

Everywhere, in field and hedgerow and garden and forest, the insect predators and parasites are at work. Here, above a pond, the dragonflies dart and the sun strikes fire from their wings. So their ancestors sped through swamps where huge reptiles lived. Now, as in those ancient times, the sharp-eyed dragonflies capture mosquitoes in the air, scooping them in with basket-shaped legs. In the waters below, their young, the dragonfly nymphs, or naiads, prey on the aquatic stages of mosquitoes and other insects.

Or there, almost invisible against a leaf, is the lacewing, with green gauze wings and golden eyes, shy and secretive, descendant of an ancient race that lived in Permian times. The adult lacewing feeds mostly on plant nectars and the honeydew of aphids, and in time she lays her eggs, each on the end of a long stalk which she fastens to a leaf. From these emerge her children—strange, bristled larvae called aphis lions, which live by preying on aphids, scales, or mites, which they capture and suck dry of fluid. Each may consume several hundred aphids before the ceaseless turning of the cycle of its life brings the time when it will spin a white silken cocoon in which to pass the pupal stage.

And there are many wasps, and flies as well, whose very existence depends on the destruction of the eggs or larvae of other insects through parasitism. Some of the egg parasites are exceedingly minute wasps, yet by their numbers and their great activity they hold down the abundance of many crop-destroying species.

All these small creatures are working—working in sun and rain, during the hours of darkness, even when winter's grip has damped down the fires of life to mere embers. Then this vital force is merely smoldering, awaiting the time to flare again into activity when spring awakens the insect world. Meanwhile, under the white blanket of snow, below the frost-hardened soil, in crevices in the bark of trees, and in sheltered caves, the parasites and the predators have found ways to tide themselves over the season of cold.

The eggs of the mantis are secure in little cases of thin parchment attached to the branch of a shrub by the mother who lived her life span with the summer that is gone.

The female *Polistes* wasp, taking shelter in a forgotten corner of some attic, carries in her body the fertilized eggs, the heritage on which the whole future of her colony depends. She, the lone survivor, will start a small paper nest in the spring, lay a few eggs in its cells, and carefully rear a small force of workers. With their help she will then enlarge the nest and develop the colony. Then the workers, foraging ceaselessly through the hot days of summer, will destroy countless caterpillars.

Thus, through the circumstances of their lives, and the nature of our own wants, all these have been our allies in keeping the balance of nature tilted in our favor. Yet we have turned our artillery against our friends. The terrible danger is that we have grossly underestimated their value in keeping at bay a dark tide of enemies that, without their help, can overrun us.

The prospect of a general and permanent lowering of environmental resistance becomes grimly and increasingly real with each passing year as the number, variety, and destructiveness of insecticides grows. With the passage of time we may expect progressively more serious outbreaks of insects, both disease-carrying and crop-destroying species, in excess of anything we have ever known.

"Yes, but isn't this all theoretical?" you may ask. "Surely it won't really happen—not in my lifetime, anyway."

But it is happening, here and now. Scientific journals had already recorded some 50 species involved in violent dislocations of nature's balance by 1958. More examples are being found every year. A recent review of the subject contained references to 215 papers reporting or discussing unfavorable upsets in the balance of insect populations caused by pesticides.

Sometimes the result of chemical spraying has been a tremendous upsurge of the very insect the spraying was intended to control, as when blackflies in Ontario became 17 times more abundant after spraying than they had been before. Or when in England an enormous outbreak of the cabbage aphid—an outbreak that had no parallel on record—followed spraying with one of the organic phosphorus chemicals.

At other times spraying, while reasonably effective against the target insect, has let loose a whole Pandora's box of destructive pests that had never previously been abundant enough to cause trouble. The spider mite, for example, has become practically a worldwide pest as DDT and other insecticides have killed off its enemies. The spider mite is not an insect. It is a barely visible eight-legged creature belonging to the group that includes spiders, scorpions, and ticks. It has mouth parts adapted for piercing and sucking, and a prodigious appetite for the chlorophyll that makes the world green. It inserts these minute and stiletto-sharp mouth parts into the outer cells of leaves and evergreen needles and extracts the chlorophyll. A mild infestation gives trees and shubbery a mottled or salt-and-pepper appearance; with a heavy mite population, foliage turns yellow and falls.

This is what happened in some of the western national forests a few years ago, when in 1956 the United States Forest Service sprayed some 885,000 acres of forested lands with DDT. The intention was to control the spruce budworm, but the following summer it was discovered that a problem worse than the budworm damage had been created. In surveying the forests from the air, vast blighted areas could be seen where the magnificent Douglas firs were turning brown and dropping their needles. In the Helena National Forest and on the western slopes of the Big Belt Mountains, then in other areas of Montana and down into Idaho the forests looked as though they had been scorched. It was evident that this summer of 1957 had brought the most extensive and spectacular infestation of spider mites in history. Almost all of the sprayed area was affected. Nowhere else was the damage evident. Searching for precedents, the foresters could remember other scourges of spider mites, though less dramatic than this one. There had been similar trouble along the Madison River in Yellowstone Park in 1929, in Colorado 20 years later, and then in New Mexico in 1956. *Each of these outbreaks had followed forest spraying with insecticides.* (The 1929 spraying, occurring before the DDT era, employed lead arsenate.)

Why does the spider mite appear to thrive on insecticides? Besides the obvious fact that it is relatively insensitive to them, there seem to be two other reasons. In nature it is kept in check by various predators such as ladybugs, a gall midge, predaceous mites and several pirate bugs, all of them extremely sensitive to insecticides. The third reason has to do with population pressure within the spider mite colonies. An undisturbed colony of mites is a densely settled community, huddled under a protective webbing for concealment from its enemies. When sprayed, the colonies disperse as the mites, irritated though not killed by the chemicals, scatter out in search of places where they will not be disturbed. In so doing they find a far greater abundance of space and food than was available in the former colonies. Their enemies are now dead so there is no need for the mites to spend their energy in secreting protective webbing. Instead, they pour all their energies into producing more mites. It is not uncommon for their egg production to be increased threefold—all through the beneficent effect of insecticides.

In the Shenandoah Valley of Virginia, a famous apple-growing region, hordes of a small insect called the red-banded leaf roller arose to plague the growers as soon as DDT began to replace arsenate of lead. Its depredations had never before been important; soon its toll rose to 50 per cent of the crop and it achieved the status of the most destructive pest of apples, not only in this region but throughout much of the East and Midwest, as the use of DDT increased.

The situation abounds in ironies. In the apple orchards of Nova Scotia in the late 1940's the worst infestations of the codling moth (cause of "wormy apples") were in the orchards regularly sprayed. In unsprayed orchards the moths were not abundant enough to cause real trouble.

Diligence in spraying had a similarly unsatisfactory reward in the eastern Sudan, where cotton growers had a bitter experience with DDT. Some 60,000 acres of cotton were being grown under irrigation in the Gash Delta. Early trials of DDT having given apparently good results, spraying was intensified. It was then that trouble began. One of the most destructive enemies of cotton is the bollworm. But the more cotton was sprayed, the more bollworms appeared. The unsprayed cotton suffered less damage to fruits and later to mature bolls than the sprayed, and in twice-sprayed fields the yield of seed cotton dropped significantly. Although some of the leaf-feeding insects were eliminated, any benefit that might thus have been gained was more than offset by bollworm damage. In the end the growers were faced with the unpleasant truth that their cotton yield would have been greater had they saved themselves the trouble and expense of spraying.

In the Belgian Congo and Uganda the results of heavy applications of DDT against an insect pest of the coffee bush were almost "catastrophic." The pest itself was found to be almost completely unaffected by the DDT, while its predator was extremely sensitive.

In America, farmers have repeatedly traded one insect enemy for a worse one as spraying upsets the population dynamics of the insect world. Two of the mass-spraying programs recently carried out have had precisely this effect. One was the fire ant eradication program in the South; the other was the spraying for the Japanese beetle in the Midwest.

When a wholesale application of heptachlor was made to the farmlands in Louisiana in 1957, the result was the unleashing of one of the worst enemies of the sugarcane crop—the sugarcane borer. Soon after the heptachlor treatment, damage by borers increased sharply. The chemical aimed at the fire ant had killed off the enemies of the borer. The crop was so severely damaged that farmers sought to bring suit against the state for negligence in not warning them that this might happen.

The same bitter lesson was learned by Illinois farmers. After the devastating bath of dieldrin recently administered to the farmlands in eastern Illinois for the control of the Japanese beetle, farmers discovered that corn borers had increased enormously in the treated area. In fact, corn grown in fields within this area contained almost twice as many of the destructive larvae of this insect as did the corn grown outside. The farmers may not yet be aware of the biological basis of what has happened, but they need no scientists to tell them they have made a poor bargain. In trying to get rid of one insect, they have brought on a scourge of a much more destructive one. According to Department of Agriculture estimates, total damage by the Japanese beetle in the United States adds up to about 10 million dollars a year, while damage by the corn borer runs to about 85 million.

It is worth noting that natural forces had been heavily relied on for control of the corn borer. Within two years after this insect was accidentally introduced from Europe in 1917, the United States Government had mounted one of its most intensive programs for locating and importing parasites of an insect pest. Since that time 24 species of parasites of the corn borer have been brought in from Europe and the Orient at considerable expense. Of these, 5 are recognized as being of distinct value in control. Needless to say, the results of all this work are now jeopardized as the enemies of the corn borer are killed off by the sprays.

If this seems absurd, consider the situation in the citrus groves of California, where the world's most famous and successful experiment in biological control was carried out in the 1880's. In 1872 a scale insect that feeds on the sap of citrus trees appeared in California and within the next 15 years developed into a pest so destructive that the fruit crop in many orchards was a complete loss. The young citrus industry was threatened with destruction. Many farmers gave up and pulled out their trees. Then a parasite of the scale insect was imported from Australia, a small lady beetle called the vedalia. Within only two years after the first shipment of the beetles, the scale was under complete control throughout the citrus-growing sections of California. From that time on one could search for days among the orange groves without finding a single scale insect.

Then in the 1940's the citrus growers began to experiment with glamorous new chemicals against other insects. With the advent of DDT and the even more toxic chemicals to follow, the populations of the vedalia in many sections of California were wiped out. Its importation had cost the government a mere $5000. Its activities had saved the fruit growers several millions of dollars a year, but in a moment of heedlessness the benefit was canceled out. Infestations of the scale insect quickly reappeared and damage exceeded anything that had been seen for fifty years.

"This possibly marked the end of an era," said Dr. Paul DeBach of the Citrus Experiment Station in Riverside. Now control of the scale has become enormously complicated. The vedalia can be maintained only by repeated releases and by the most careful attention to spray schedules, to minimize their contact with insecticides. And regardless of what the citrus growers do, they are more or less at the mercy of the owners of adjacent acreages, for severe damage has been done by insecticidal drift.

All these examples concern insects that attack agricultural crops. What of those that carry disease? There have already been warnings. On Nissan Island in the South Pacific, for example, spraying had been carried on intensively during the Second World War, but was stopped when hostilities came to an end. Soon swarms of a malaria-carrying mosquito reinvaded the island. All of its predators had been killed off and there had not been time for new populations to become established. The way was therefore clear for a tremendous population explosion. Marshall Laird, who has described this incident, compares chemical control to a treadmill; once we have set foot on it we are unable to stop for fear of the consequences.

In some parts of the world disease can be linked with spraying in quite a different way. For some reason, snail-like mollusks seem to be almost immune to the effects of insecticides. This has been observed many times. In the general holocaust that followed the spraying of salt marshes in eastern Florida, . . . aquatic snails alone survived. The scene as described was a macabre picture—something that might have been created by a surrealist brush. The snails moved among the bodies of the dead fishes and the moribund crabs, devouring the victims of the death rain of poison.

But why is this important? It is important because many aquatic snails serve as hosts of dangerous parasitic worms that spend part of their life cycle in a mollusk, part in a human being. Examples are the blood flukes, or schistosoma, that cause serious disease in man when they enter the body by way of drinking water or through the skin when people are bathing in infested waters. The flukes are released into the water by the host snails. Such diseases are especially prevalent in parts of Asia and Africa. Where they occur, insect control measures that favor a vast increase of snails are likely to be followed by grave consequences.

And of course man is not alone in being subject to snail-borne disease. Liver disease in cattle, sheep, goats, deer, elk, rabbits, and various other warm-blooded animals may be caused by liver flukes that spend part of their life cycles in fresh-water snails. Livers infested with these worms are unfit for use as human food and are routinely condemned. Such rejections cost American cattlemen about 3½ million dollars annually. Anything that acts to increase the number of snails can obviously make this problem an even more serious one.

Over the past decade these problems have cast long shadows, but we have been slow to recognize them. Most of those best fitted to develop natural controls and assist in putting them into effect have been too busy laboring in the more exciting vineyards of chemical control. It was reported in 1960 that only 2 per cent of all the economic entomologists in the country were then working in the field of biological controls. A substantial number of the remaining 98 per cent were engaged in research on chemical insecticides.

Why should this be? The major chemical companies are pouring money into the universities to support research on insecticides. This creates attractive fellowships for graduate students and attractive staff positions. Biological-control studies, on the other hand, are never so endowed—for the simple reason that they do not promise anyone the fortunes that are to be made in the chemical industry. These are left to state and federal agencies, where the salaries paid are far less.

This situation also explains the otherwise mystifying fact that certain outstanding entomologists are among the leading advocates of chemical control. Inquiry into the background of some of these men reveals that their entire research program is supported by the chemical industry. Their professional prestige, sometimes their very jobs depend on the perpetuation of chemical methods. Can we then expect them to bite the hand that literally feeds them? But knowing their bias, how much credence can we give to their protests that insecticides are harmless?

Amid the general acclaim for chemicals as the principal method of insect control, minority reports have occasionally been filed by those few entomologists who have not lost sight of the fact that they are neither chemists nor engineers, but biologists.

F. H. Jacob in England has declared that "the activities of many so-called economic entomologists would make it appear that they operate in the belief that salvation lies at the end of a spray nozzle . . . that when they have created problems of resurgence or resistance or mammalian toxicity, the chemist will be ready with another pill. That view is not held here. . . . Ultimately only the biologist will provide the answers to the basic problems of pest control."

"Economic entomologists must realize," wrote A. D. Pickett of Nova Scotia, "that they are dealing with living things . . . their work must be more than simply insecticide testing or a quest for highly destructive chemicals." Dr. Pickett himself was a pioneer in the field of working out sane methods of insect control that take full advantage of the predatory and parasitic species. The method which he and his associates evolved is today a shining model but one too little emulated. Only in the integrated control programs developed by some California entomologists do we find anything comparable in this country.

Dr. Pickett began his work some thirty-five years ago in the apple orchards of the Annapolis Valley in Nova Scotia, once one of the most concentrated fruit-growing areas in Canada. At that time it was believed that insecticides—then inorganic chemicals—would solve the problems of insect control, that the only task was to induce fruit growers to follow the recommended methods. But the rosy picture failed to materialize. Somehow the insects persisted. New chemicals were added, better spraying equipment was devised, and the zeal for spraying increased, but the insect problem did not get any better. Then DDT promised to "obliterate the nightmare" of codling moth outbreaks. What actually resulted from its use was an unprecedented scourge of mites. "We move from crisis to crisis, merely trading one problem for another," said Dr. Pickett.

At this point, however, Dr. Pickett and his associates struck out on a new road instead of going along with other entomologists who continued to pursue the will-o'-the-wisp of the ever more toxic chemical. Recognizing that they had a strong ally in nature, they devised a program that makes maximum use of natural controls and minimum use of insecticides. Whenever insecticides are applied only minimum dosages are used—barely enough to control the pest without avoidable harm to beneficial species. Proper timing also enters in. Thus, if nicotine sulphate is applied before rather than after the apple blossoms turn pink one of the important predators is spared, probably because it is still in the egg stage.

Dr. Pickett uses special care to select chemicals that will do as little harm as possible to insect parasites and predators. "When we reach the point of using DDT, parathion, chlordane, and other new insecticides as routine control measures in the same way we have used the inorganic chemicals in the past, entomologists interested in biological control may as well throw in the sponge," he says. Instead of these highly toxic, broad-spectrum insecticides, he places chief reliance on ryania (derived from ground stems of a tropical plant), nocotine sulphate, and lead arsenate. In certain situations very weak concentrations of DDT or malathion are used (1 or 2 ounces per 100 gallons—in contrast to the usual 1 or 2 pounds per 100 gallons). Although these two are the least toxic of the modern insecticides, Dr. Pickett hopes by further research to replace them with safer and more selective materials.

How well has this program worked? Nova Scotia orchardists who are following Dr. Pickett's modified spray program are producing as high a proportion of first-grade fruit as are those who are using intensive chemical applications. They are also getting as good production. They are getting these results, moreover, at a substantially lower cost. The outlay for insecticides in Nova Scotia apple orchards is only from 10 to 20 per cent of the amount spent in most other apple-growing areas.

More important than even these excellent results is the fact that the modified program worked out by these Nova Scotian entomologists is not doing violence to nature's balance. It is well on the way to realizing the philosophy stated by the Canadian entomologist G. C. Ullyett a decade ago: "We must change our philosophy, abandon our attitude of human superiority and admit that in many cases in natural environments we find ways and means of limiting populations of organisms in a more economical way than we can do it ourselves."

INTRODUCTION

If man is to save his world, he must know himself and what he has done and is doing to his world. In Part 1, we looked at what man has done to his world. In Part 2, we will look at that world as it should be, and in many places still is.

For some strange reason, man still pictures himself at the center of his world. The reality is that he is only one kind of animal living with millions of other kinds of animals and plants; one kind of creature with a big brain, surrounded by as many planets and stars as there are grains of sand on all the beaches of the world. While we cannot at present tell whether life exists on other planets, we can study life on our mother planet, earth.

The province of the biologist is to do exactly that. In "Log from the Sea of Cortez," John Steinbeck, the Nobel Prize-winning novelist, says,

> What good men most biologists are, the tenors of the scientific world—temperamental, moody, loud-laughing and healthy. . . . The true biologist deals with life, with teeming boisterous life, and learns something from it, learns that the first rule of life is living.

As a biologist-naturalist, I am flattered by Steinbeck's evaluation of my colleagues and myself. But as a teaching biologist, I know that not everyone shares my reverence for life.

In his remarkable book, "The Forest and the Sea," Marston Bates remarks,

> People often come to me with some strange animal they have found. "What is it?", they ask. "Oh," I say brightly, "that is a swallow tail butterfly, *Paoilio cresphontes.*" It is curious how happy people are to have a name for something, for an animal or plant, even though they know nothing about it beyond the name. But other questions follow, "Where does it live?" and "What does it do?" . . . Almost inevitably there will come another question, "What good is it?" Faced with astronomical space and geological time, faced with the immense diversity of living forms, how can one ask of one particular kind of butterfly, "What good is it?" Often my reaction is to ask in turn, "What good are you?"

As a teacher, I have run into these same questions on numerous occasions. I also am at a loss to answer the question, "What good is it?" And I fear that there is little hope for man as an organism crawling through the landscape unless there is a resurgence of feeling for himself as one, just one, of all the living creatures which still populate the earth.

It is not the man who loves the natural world who asks, "What good is it?" A brief study of natural biology can expose you to a few out-of-doors experiences in nature and so possibly awaken or reawaken your almost instinctual need to feel at one with her. And that feeling cannot be quantitatively evaluated, as we can evaluate the damage man is doing to his world. What instruments has our computerized society devised which can measure and chart the feelings of a teacher and a handful of interested students when they are reverently peeping into the miniature world of the crystal-clear tide pool? How can you measure the visual impact of the sea urchin's deep purple spines, the sunny orange of the encrusting sponge, the living sea-green tentacles of the sea anemone, the lacy fragility of the red sea flower? How does one measure one's feelings while walking under the towering spires of California's redwoods or when desert wildflowers carpet the sand in a blinding profusion of color?

Learn to live in the world of nature. As Steinbeck said,

Let's go wide open. Let's see what we see, record what we find, and not fool ourselves with conventional scientific structure. . . . Let us go, into the Sea of Cortez, realizing that we become forever a part of it: that our rubber boots slogging through a flat of eelgrass, that the rocks we turn over, makes us truly a factor in the ecology of the region. We shall take something away from it but we will leave something too. And if we seem a small factor in a huge pattern, nevertheless it is of relative importance.

DON'T EXPECT TOO MUCH FROM A FROG *

JOSEPH WOOD KRUTCH

Joseph Wood Krutch was one of those rare human beings who occasionally steps on the stage of life. He was a drama critic, teacher, naturalist, philosopher, and man of letters.

In this essay, *Don't Expect Too Much From a Frog*, you catch a glimpse of the marvels and comedies of nature. Professor Krutch always searched for a common meeting ground with other living things. He said, in fact, "I always feel more serene after a conversation with a few friendly animals than I do after an evening with even the most brilliant of my human acquaintances."

Some of his books include "Henry David Thoreau," "The Twelve Seasons," and "The Desert Year."

* From "The Best of Two Worlds" by Joseph Wood Krutch. Published by William Sloane Associates. Reprinted by permission of William Morrow & Company, Inc. Copyright © 1950, 1951, 1953 by Joseph Wood Krutch. (This essay was first published in *The Southwest Review*.)

Whenever I feel that I would like to see a frog—and you would be surprised how often that is—I have only to open the door of my living room and take two or three steps to a small pool in the shade of a large spiraea.

Before the fox got my two pet ducks that wintry day a few years ago this pool was theirs. For a full decade they swam gravely about on its six-by-ten foot surface, blissfully happy and apparently unaware that water ever comes in larger pieces. They also chased one another with dizzy playfulness around and around its narrow circumference, stood on their heads to search for the worms they never found on its concrete bottom, and splashed madly at least once a day for the shower baths they love. From those ducks in their time I got many lessons in gladness and much moral instruction: A little world, I learned, is as big as one thinks and makes it. To be a big duck in a small pond is not necessarily ignoble. One can have Lake Michigan in one's back yard if one wants to. Water is much the same everywhere. It is better to use what one has than to regret what one has not. Where ignorance is bliss. . . . If we are unhappy it is less often because of something we lack than because we do not know how to use what we have. Ducks want but little here below.

When, after a long and happy life—one had become a mother for the first time at the age of twelve years—these ducks at last fed the fox, I decided to keep frogs. I refurbished the pool which had leaked and gone dry, and because I am, I fear, a man of too little faith even in nature, I planned vaguely to collect from some pond specimens which might be persuaded to settle down with me.

I should have known better. Two days after my pool had been filled the first frog had discovered it and taken up permanent residence. Within a week there was at least one individual of the four usual species of this region: the two spotted kinds, Palustris and Pipiens, which most people don't bother to tell apart; Clamitans, the robust fellow with the green neck and head; and Catesbiana, the huge basso profundo who alone has a right to the name "bullfrog." Sometimes these frogs now hide in the herbage or under a stone. Nearly always, however, at least one or two sit on the rough, slab-topped rim of the pool—motionless, unblinking, sublimely confident that sooner or later, but always quite safely in time, some insect will pass by to be snatched up with a lightning tongue. Its unerring, flashing speed is proof enough that the frogs are not unaware of the external world as they sit, seemingly absorbed in meditation, hour after hour. Yet for all that, they are as exempt as the lily from the curse of Adam. They do not work for a living. They do not hunt for food. They do not even sing for their supper. They merely wait for it to pass by and sing afterward.

As the summer rolls along I hope that I may learn something from them as I did from their feathered predecessors. But whatever they may have to teach me will not be as easy to translate into human terms as the lesson of the ducks. Because these last were warm-blooded they were much closer to me in situation and philosophy. Gladness and pain were for us recognizably the same things. They could set me a real example when they rejoiced so unmistakably in the little pool which had to serve them in lieu of a lake. But the frogs are antediluvian. Like me, they are, to be sure, alive, and their protoplasm is much like mine. But they seem strangely cold-blooded in a figurative as well as a literal sense. No doubt they feel the difference between well-being and its reverse; their satisfactions may be, for all I know, as deep or deeper than mine. But their monumental placidity is something which I can hardly hope to achieve.

That is one of the reasons why I like to have them about, why I am so acutely aware of the strangeness of having almost at my doorstep these creatures who can demonstrate what to be alive must have meant millions of years ago when the Amphibia were the most progressive, the most adventurous, one is almost tempted to say the most warm-blooded and passionate, of living creatures.

I need do no more than notice the grasshopper just fallen into the pool to realize that, comparatively speaking, the frogs are more like me than I had supposed. At least neither their anatomy nor their ways are as remote from mine as are those of the insects whose tastes and habits sometimes shocked even Henri Fabre with their monstrousness. So far as remoteness of soul is concerned there is at least as much difference between the grasshopper and the frog as between the frog and the duck; perhaps indeed more than between the duck and me. The frog got to be what he is by moving in the direction which was to end—for the time being at least—in me. Even though land insects were newcomers when the frogs were already a well established family they took long ago the direction which was to carry their descendants further and further away from the rest of us; further and further from the frog, the duck, and the man. When I realize that, I feel that Catesbiana and I are not so different, as differences go in this world, as I had thought.

At least one thing which I already knew the frogs have by now taught several visitors who consented to inspect them. Part-time countrymen though all these visitors were, several saw for the first time that the frog, like the toad, is not "ugly and venomous," but "interesting looking." And this I know is halfway to the truth that they are beautiful in their own fashion and ask only that we enlarge our conception of beauty to include one more of nature's many kinds. They are as triumphantly *what* they are as man has ever succeeded in being, and no human imagination could devise a new creature so completely "right" according to the laws of his own structure and being. They are the perfect embodiment of frogginess both outwardly and inwardly, so that nearly every individual achieves, as very few human beings do, what the Greeks would have called his entelechy—the complete realization of the possibilities which he suggests. Frogs are not shadows of an ideal but the ideal itself. There is scarcely one who is not either a froggy Belvedere or a froggy de Milo. To realize this one need only go to frogs as so many critics have told us we should go to works of art, asking first "What is the intention" and then "How well has this intention been achieved?"

I do not know what human being was the first to say that reptiles are not "vile" and toads not "loathsome." I do know that these clichés lingered even in the literature of natural history almost down to our own time and that perhaps most people still pick their way foppishly about the world, rather proud of the distaste aroused by almost everything which shares the world with them and apparently convinced that nearly everything except themselves was a mistake on the part of the Creator. Even Linnaeus expressed a distaste for the Amphibia and the reptiles. Even Gilbert White, who certainly knew better and felt better, absentmindedly falls into the literary convention of his day and once refers to the ancient tortoise in his garden as a "vile reptile." Most moderns are still far from catching up with Sir Thomas Browne who, three hundred years ago, wrote into his *Religio Medici* something for which I am very grateful: "I cannot tell by what logic we call a toad, a bear or an elephant ugly; they being created in those outward shapes and figures which best express the actions of their inward forms. . . . To speak yet more narrowly, there was never anything ugly or misshapen, but Chaos. . . . All things are artificial; for Nature is the Art of God."

One of my lady visitors after looking for a while at Catesbiana expressed the opinion that he was "cute," and I restrained myself because I recognized that she meant well, that this was even, for her, the first faint beginning of wisdom. I did not ask as I was about to ask if the prehistoric saurians were "cute," if the stretches of geologic time are "nice," and if the forests of the carboniferous age are "sweet." As my frog gazed at both of us with eyes which had seen the first mammalian dawn, lived through thousands of millennia before the first lowly primate appeared and so saw more of the world than all mankind put together, I asked him to forgive the impertinence of this latecomer. The Orientals say that God made the cat in order that man might have the pleasure of petting a tiger. By that logic He must have made frogs so that we can commune with prehistory.

During the two centuries just past more has been written about the attitude of man toward the animals than in the two millennia before. Very little has been said even yet about their attitude toward us, and I sometimes wonder if that is not a coming topic just as its converse was beginning to be when Alexander Pope wrote his *Essay on Man*—which is, as a matter of fact, almost as much about man's fellow creatures as it is about man himself. At least until dogs and cats begin to write books there is bound to be a rather large element of speculation in any treatment of the subject, but we can imagine what some of the domestic animals might say, as Miss Sylvia Townsend Warner did when she made one disillusioned canine say to another: "The more I see of dogs the better I like people." That, I am afraid, is a bit anthropomorphic and so perhaps are the implications of my own frequent wish that I could hear the comments of, say, some wise, tolerant collie on the conclusions concerning human nature to be drawn from the juxtaposition of the statement that "the dog is man's best friend" with the familiar exclamation, "I wouldn't treat a dog that way."

It is not, however, in the spirit of either frivolity or cynicism that I wonder what my frogs think of me. They think something because they show some awareness of my existence and the nature of that awareness, like the nature of the other relations between us, shows again how much more remote they are than the ducks, how much less remote than certain other creatures who are, nevertheless, in some sense alive. It is hard to believe that the insects are aware that we exist at all except as an impersonal force. In some part that may be because, with one exception, they cannot turn their heads over their shoulders to look at us. The exception is the praying mantis and when he cocks his improbable head over his improbable shoulder, I have to remind myself that he is one of the most primitive insects and not, as this gesture might suggest, the most like one of us. But I doubt that he, or a grasshopper, or even a bee, takes me in as, in some realer sense, the frog quite obviously does when he considers whether or not he had better leap into the water at my approach and comes to do so less and less readily as he gets to know me better. The difference between his awareness and that of a cat or dog who can understand even some of our words is immense, but perhaps no more so than the difference between that of the frog and the apparently utter obliviousness of the insect who is shut off not merely because he is less intelligent but also because such intelligence as he has is so utterly different from ours. His standards of value, if I may put it that way, are too irreconcilable with mine, with the cat's and with the frog's.

On a terrarium by my window is a huge ten-inch salamander from Florida with whom I have had some relation for eight or nine years. At least when he is hungry he looks up if I happen to pass by and he will waddle toward me if I offer a worm. Already my bullfrog will take beef from the end of a string when I offer it, but he has given no sign that he connects me with the bounty, and if he comes at last to do so, I shall still not know whether it is more than a quasi-mechanical association of a sequence of events. My old housekeeper repeatedly assures me that the salamander "knows her," and though I think that this is probably a pretty large overinterpretation, I am not convinced that the animal mind is as nearly a mere set of reflexes as the behaviorists confidently assert.

I admit that in eight or nine years this salamander and I have got into no very intimate, two-sided relationship, and this makes me wonder how much the budding acquaintance between the frog and me will come to. On his side, I imagine, it will not come to very much; I will not mean much in his consciousness. He is nearer to me than a grasshopper would be but nearer only as the sun is nearer than Betelgeuse or Betelgeuse is nearer than the closest of the extragalactic nebulae. The distance is still vast as we measure things. We can look at one another as the insect and I probably cannot, but we look across a gap as wide as that which separates the frog's origin in impossibly remote time from mine only a few thousand years ago. Perhaps some physicist who likes to think of time in terms of arrows which may point in one direction while other time-arrows point in the opposite, would say that the gap across which we look actually is a temporal one. I am looking back at him and he is looking forward toward me—across the interval which separates the ancient time when he was the latest model from this present time when I am.

Perhaps I am already beginning to be an unimportant part of his dim consciousness. But whatever intercourse may ultimately develop between us will be marked by considerable reserve on his part, and he will never want to climb into my lap as a certain goose used to do. A calm sort of mutual respect is the most I look forward to. Of one thing, however, I am sure from experience: whatever relationship we do achieve will be on the basis of reciprocal tolerance and good manners. The lower animals seldom bicker and at most usually suggest only that we should leave them alone. They may fight for self-protection if one insists, but they would prefer only to agree to differ. When my frog gets enough of my company he will show me a fine pair of hind legs, but he will hurl no insults and make no cutting remarks. He will say only, "I should prefer to be alone."

Perhaps this is partly because he can make a cleaner getaway than people can when someone bores them, but I have noticed also and in general that only the higher animals can be bad tempered. I have seen a seal deliberately squirt a stream of water into a spectator's face—in fact the New York Aquarium had, many years ago, a notorious member of that species who did this regularly to the great delight of those waiting spectators who were in the know. But that was only rough humor and I have never been deliberately insulted by any creature lower than a chimpanzee. That was when I set one such an example of bad manners by deliberately trying to stare him down as he looked from behind his bars. Presently, he turned his back to me with great deliberation, lowered his head until he was peering out from between his buttocks, and then pursed his lips to give me what is commonly known as a Bronx Cheer, though the present use of it seemed to suggest that it had been invented long before the borough was founded. I have never seen any other subhuman primate do anything which seemed so completely human, and I never at any moment in my life had a more lively conviction that we and the anthropoids really do come of a common stock.

My intercourse with the frogs will never be marked by any incidents like that. Indeed it will probably seldom provide what could be called an incident at all. But that is the price one must pay for escaping also the possibility of anything approaching a disturbing clash. His sang-froid is what the words mean literally: cold blood. My blood is too warm not to want, often, a more lively if more dangerous inter-course—with ducks, with cats, and even with creatures as exasperating as the members of my own species frequently are.

Much the same sort of thing I am compelled to admit in connection with the frog's vocal comments. The four species gathered around my pool—which of course they legitimately think of as theirs—are respectively soprano, tenor, contralto, and bass. But what they croak at their various pitches is the same thing, even though it sounds weightiest when Catesbiana gives it utterance. The whole philosophy of frogs, all the wisdom they have accumulated in millions of years of experience, is expressed in that *urrrr-unk* uttered with an air which seems to suggest that the speaker feels it to be completely adequate. The comment does not seem very passionate or very aspiring, but it is contented and not cynical. Frogs have considered life and found it, if not exactly ecstatic, at least quite pleasant and satisfactory. Buddha is said to have made a comment much like theirs after all those years of contemplating his navel, and though I do not wholly understand it I think I catch the drift.

It is not, I realize, quite enough for me whether it comes from the pond-side or from the Mysterious East. I cannot live by it all the time. Not infrequently I find myself responding instead to the more passionate observations of Shakespeare or Mozart. But in that eclecticism which serves me as a philosophy of life it has its place. It is basic and a last resort. I like to hear it preached from time to time and to think that it is something one might at least fall back upon.

BIRTH OF A MACKEREL*

RACHEL L. CARSON

In "Under the Sea-Wind," with her gifted scientific prose, Rachel Carson wrote about the life cycle of Scomber, the mackerel. Scomber's life and his survival contains all the thrills of the chase, enacted among mysterious and sometimes terrifying forms far below the surface of the sea.

Rachel Carson's other books include her National Book Award-winning "The Sea Around Us." Her beautifully written "Edge of the Sea" and "Silent Spring" are both classics of their kind.

o it came about that Scomber, the mackerel, was born in the surface waters of the open sea, seventy miles south by east from the western tip of Long Island. He came into being as a tiny globule no larger than a poppy seed, drifting in the surface layers of pale-green water. The globule carried an amber droplet of oil that served to keep it afloat and it carried also a gray particle of living matter so small that it could have been picked up on the point of a needle. In time this particle was to become Scomber, the mackerel, a powerful fish, streamlined after the manner of his kind, and a rover of the seas.

The parents of Scomber were fish of the last big wave of mackerel migration that came in from the edge of the continental shelf in May, heavy with spawn and driving rapidly shoreward. On the fourth evening of their journey, in a flooding current straining to landward, the eggs and milt had begun to flow from their bodies into the sea. Somewhere among the forty or fifty thousand eggs that were shed by one of the female fish was the egg that was to become Scomber.

There could be scarcely a stranger place in the world in which to begin life than this universe of sky and water, peopled by strange creatures and governed by wind and sun and ocean currents. It was a place of silence, except when the wind went whispering or blustering over the vast sheet of water, or when sea gulls came down the wind with their high, wild mewing, or when whales broke the surface, expelled the long-held breath, and rolled again into the sea.

The mackerel schools hurried on into the north and east, their journey scarcely interrupted by the act of spawning. As the sea birds were finding their resting places for the night on the dark water plains, swarms of small and curiously formed animals stole into the surface waters from hills and valleys lying in darkness far below. The night sea belonged to the plankton, to the diminutive worms and the baby crabs, the glassy, big-eyed shrimp, the young barnacles and mussels, the throbbing bells of the jellyfish, and all the other small fry of the sea that shun the light.

It was indeed a strange world in which to set adrift anything so fragile as a mackerel egg. It was filled with small hunters, each of which must live at the expense of its neighbors, plant and animal. The eggs of the mackerel were jostled by the newly hatched young of earlier spawning fishes and of shellfish, crustaceans, and worms. The larvae, some of them only a few hours old, were swimming alone in the sea, busily seeking their food. Some snatched out of the water with pincered claws anything small enough to be overpowered and swallowed; others seized any prey less swift and agile than themselves in biting jaws or sucked into cilium-studded mouths the drifting green or golden cells of the diatoms.

The sea was filled, too, with larger hunters than the microscopic larvae. Within an hour after the parent mackerel had gone away, a horde of comb jellies rose to the surface of the sea. The comb jellies, or ctenophores, looked like large gooseberries, and they swam by the beating of plates of fused hairs or cilia, set in eight bands down the sides of the transparent bodies. Their substance was scarcely more than that of sea water, yet each of them ate many times its own bulk of solid food in a day. Now they were rising slowly toward the surface, where the millions of new-spawned mackerel eggs drifted free in the upper layers of the sea. They twirled slowly back and forth on the long axes of their bodies as they came, flashing a cold, phosphorescent fire. Throughout the night the ctenophores flicked the waters with their deadly tentacles, each a slim, elastic thread twenty times the length of the body when extended. And as they turned and twirled and flashed frosty green lights in the black water, jostling one another in their greed, the drifting mackerel eggs were swept up in the silken meshes of the tentacles and carried by swift contraction to the waiting mouths.

Often during this first night of Scomber's existence the cold, smooth body of a ctenophore collided with him or a searching tentacle missed by a fraction of an inch the floating sphere in which the speck of protoplasm had already divided into eight parts, thus beginning the development by which a single fertile cell would swiftly be transformed into an embryo fish.

Of the millions of mackerel eggs drifting alongside the one that was to produce Scomber, thousands went no farther than the first stages of the journey into life until they were seized and eaten by the comb jellies, to be speedily converted into the watery tissue of their foe and in this reincarnation to roam the sea, preying on their own kind.

Throughout the night, while the sea lay under a windless sky, the decimation of the mackerel eggs continued. Shortly before dawn the water began to stir to a breeze from the east and in an hour was rolling heavily under a wind that blew steadily to the south and west. At the first ruffling of the surface calm the comb jellies began to sink into deep water. Even in these simple creatures, which consist of little more than two layers of cells, one inside the other, there exists the counterpart of an instinct of self-preservation, causing them in some way to sense the threat of destruction which rough water holds for so fragile a body.

In the first night of their existence more than ten out of every hundred mackerel eggs either had been eaten by the comb jellies or, from some inherent weakness, had died after the first few divisions of the cell.

Now, the rising up of a strong wind blowing to southward brought fresh dangers to the mackerel eggs, left for the time being with few enemies in the surface waters about them. The upper layers of the sea streamed in the direction urged upon them by the wind. The drifting spheres moved south and west with the current, for the eggs of all sea creatures are carried helplessly wherever the sea takes them. It happened that the southwest drift of the water was carrying the mackerel eggs away from the normal nursery grounds of their kind into waters where food for young fish was scarce and hungry predators abundant. As a result of this mischance fewer than one egg in every thousand was to complete its development.

On the second day, as the cells within the golden globules of the eggs multiplied by countless divisions, and the shieldlike forms of embryo fish began to take shape above the yolk spheres, hordes of a new enemy came roving through the drifting plankton. The glassworms were transparent and slender creatures that cleaved the water like arrows, darting in all directions to seize fish eggs, copepods, and even others of their own kind. With their fierce heads and toothed jaws they were terrible as dragons to the smaller beings of the plankton, although as men measure they were less than a quarter of an inch long.

The floating mackerel eggs were scattered and buffeted by the dartings and rushes of the glassworms, and when the driftings of current and tide carried them away to other waters a heavy toll of the mackerel had been taken as food.

Again the egg that contained the embryonic Scomber had drifted unscathed while all about him other eggs had been seized and eaten. Under the warm May sun the new young cells of the egg were stirred to furious activity—growing, dividing, differentiating into cell layers and tissues and organs. After two nights and two days of life, the threadlike body of a fish was taking form within the egg, curled halfway around the globe of yolk that gave it food. Already a thin ridge down the mid-line showed where a stiffening rod of cartilage—forerunner of a backbone—was forming; a large bulge at the forward end showed the place of the head, and on it two smaller outpushings marked the future eyes of Scomber. On the third day a dozen V-shaped plates of muscle were marked out on either side of the backbone; the lobes of the brain showed through the still-transparent tissues of the head; the ear sacs appeared; the eyes neared completion and showed dark through the egg wall, peering sightlessly into the surrounding world of the sea. As the sky lightened preparatory to the fifth rising of the sun a thin-walled sac beneath the head—crimson tinted from the fluid it contained— quivered, throbbed, and began the steady pulsation that would continue as long as there was life within the body of Scomber.

Throughout that day development proceeded at a furious pace, as though in haste to make ready for the hatching that was soon to come. On the lengthening tail a thin flange of tissue appeared—the fin ridge from which a series of tail finlets, like a row of flags stiff in the wind, was later to be formed. The sides of an open groove that traversed the belly of the little fish, beneath and protected by the plate of more than seventy muscle segments, grew steadily downward and in mid-afternoon closed to form the alimentary canal. Above the pulsating heart the mouth cavity deepened, but it was still far short of reaching the canal.

Throughout all this time the surface currents of the sea were pouring steadily to the southwest, driven by the wind and carrying with them the clouds of plankton. During the six days since the spawning of the mackerel the toll of the ocean's predators had continued without abatement, so that already more than half of the eggs had been eaten or had died in development.

It was the nights that had seen the greatest destruction. They had been dark nights with the sea lying calm under a wide sky. On those nights the little stars of the plankton had rivaled in number and brilliance the constellations of the sky. From underlying depths the hordes of comb jellies and glassworms, copepods and shrimps, medusae of jellyfish, and translucent winged snails had risen into the upper layers to glitter in the dark water.

When the first dilution of blackness came in the east, warning of the dawn into which the revolving earth was carrying them, strange processions began to hurry down through the water as the animals of the plankton fled from the sun that had not yet risen. Only a few of these small creatures could endure the surface waters by day except when clouds deflected the fierce lances of the sun.

In time Scomber and the other baby mackerel would join the hurrying caravans that moved down into deep green water by day and pressed upward again as the earth swung once more into darkness. Now, while still confined within the egg, the embryonic mackerel had no power of independent motion, for the eggs remained in water of a density equal to their own and were carried horizontally in their own stratum of the sea.

On the sixth day the currents took the mackerel eggs over a large shoal thickly populated with crabs. It was the spawning season of the crabs—the time when the eggs, that had been carried throughout the winter by the females, burst their shells and released the small, goblin-like larvae. Without delay the crab larvae set out for the upper waters, where through successive moltings of their infant shells and transformations of appearance they would take on the form of their race. Only after a period of life in the plankton would they be admitted to the colony of crabs that lived on that pleasant undersea plateau.

Now they hastened upward, each newborn crab swimming steadily with its wandlike appendages, each ready to discern with large black eyes and to seize with sharp-beaked mouth such food as the sea might offer. For the rest of that day the crab larvae were carried along with the mackerel eggs, on which they fed heavily. In the evening the struggle of two currents—the tidal current and the wind-driven current—carried many of the crab larvae to landward while the mackerel eggs continued to the south.

There were many signs in the sea of the approach to more southern latitudes. The night before the appearance of the crab larvae the sea had been set aglitter over an area of many miles with the intense green lights of the southern comb jelly Mnemiopsis, whose ciliated combs gleam with the colors of the rainbow by day and sparkle like emeralds in the night sea. And now for the first time there throbbed in the warm surface waters the pale southern form of the jellyfish Cyanea, trailing its several hundred tentacles through the water for fish or whatever else it might entangle. For hours at a time the ocean seethed with great shoals of salpae—thimble-sized, transparent barrels hooped in strands of muscle.

On the sixth night after the spawning of the mackerel the tough little skins of the eggs began to burst. One by one the tiny fishlets, so small that the combined length of twenty of them, head to tail, would have been scarcely an inch, slipped out of the confining spheres and knew for the first time the touch of the sea. Among these hatching fish was Scomber.

He was obviously an unfinished little fish. It seemed almost that he had burst prematurely from the egg, so unready was he to care for himself. The gill slits were marked out but were not cut through to the throat, so were useless for breathing. His mouth was only a blind sac. Fortunately for the newly hatched fishlet, a supply of food remained in the yolk sac still attached to him, and on this he would live until his mouth was open and functioning. Because of the bulky sac, however, the baby mackerel drifted upside down in the water, helpless to control his movements.

The next three days of life brought startling transformations. As the processes of development forged onward, the mouth and gill structures were completed and the finlets sprouting from back and sides and underparts grew and found strength and certainty of movement. The eyes became deep blue with pigment, and now it may be that they sent to the tiny brain the first messages of things seen. Steadily the yolk mass shrank, and with its loss Scomber found it possible to right himself and by undulation of the still-rotund body and movement of the fins to swim through the water.

Of the steady drift, the southward pouring of the water day after day, he was unconscious, but the feeble strength of his fins was no match for the currents. He floated where the sea carried him, now a rightful member of the drifting community of the plankton.

THE BEE-PASTURES *

JOHN MUIR

John Muir, Scottish-born American naturalist, lived and worked chiefly in Yosemite Valley in California. He was an active and eloquent defender of wilderness areas. He also worked diligently to help establish America's excellent National Park System.

John Muir was a rugged outdoorsman, but his writings reflect his sensitivity for the beauties of nature and a profound respect for life. His books include "The Mountains of California," "Our National Parks," and "The Yosemite."

*From "The Mountains of California" by John Muir. Copyright © 1894, 1911 by The Century Co., affiliate of Meredith Press.

When California was wild, it was one sweet bee-garden throughout its entire length, north and south, and all the way across from the snowy Sierra to the ocean.

Wherever a bee might fly within the bounds of this virgin wilderness—through the redwood forests, along the banks of the rivers, along the bluffs and headlands fronting the sea, over valley and plain, park and grove, and deep, leafy glen, or far up the piny slopes of the mountains—throughout every belt and section of climate up to the timber line, bee-flowers bloomed in lavish abundance. Here they grew more or less apart in special sheets and patches of no great size, there in broad, flowing folds hundreds of miles in length—zones of polleny forests, zones of flowery chaparral, stream-tangles of rubus and wild rose, sheets of golden compositae, beds of violets, beds of mint, beds of bryanthus and clover, and so on, certain species blooming somewhere all the year round.

But of late years plows and sheep have made sad havoc in these glorious pastures, destroying tens of thousands of the flowery acres like a fire, and banishing many species of the best honey-plants to rocky cliffs and fence-corners, while, on the other hand, cultivation thus far has given no adequate compensation, at least in kind; only acres of alfalfa for miles of the richest wild pasture, ornamental roses and honeysuckles around cottage doors for cascades of wild roses in the dells, and small, square orchards and orange-groves for broad mountain-belts of chaparral.

The Great Central Plain of California, during the months of March, April, and May, was one smooth, continuous bed of honey-bloom, so marvelously rich that, in walking from one end of it to the other, a distance of more than 400 miles, your foot would press about a hundred flowers at every step. Mints, gilias, nemophilas, castilleias, and innumerable compositae were so crowded together that, had ninety-nine per cent. of them been taken away, the plain would still have seemed to any but Californians extravagantly flowery. The radiant, honeyful corollas, touching and overlapping, and rising above one another, glowed in the living light like a sunset sky—one sheet of purple and gold, with the bright Sacramento pouring through the midst of it from the north, the San Joaquin from the south, and their many tributaries sweeping in at right angles from the mountains, dividing the plain into sections fringed with trees.

Along the rivers there is a strip of bottom-land, countersunk beneath the general level, and wider toward the foot-hills, where magnificent oaks, from three to eight feet in diameter, cast grateful masses of shade over the open, prairie-like levels. And close along the water's edge there was a fine jungle of tropical luxuriance, composed of wild-rose and bramble bushes and a great variety of climbing vines, wreathing and interlacing the branches and trunks of willows and alders, and swinging across from summit to summit in heavy festoons. Here the wild bees reveled in fresh bloom long after the flowers of the drier plain had withered and gone to seed. And in midsummer, when the "blackberries" were ripe, the Indians came from the mountains to feast—men, women, and babies in long, noisy trains, often joined by the farmers of the neighborhood, who gathered this wild fruit with commendable appreciation of its superior flavor, while their home orchards were full of ripe peaches, apricots, nectarines, and figs, and their vineyards were laden with grapes. But, though these luxuriant, shaggy river-beds were thus distinct from the smooth, treeless plain, they made no heavy dividing lines in general views. The whole appeared as one continuous sheet of bloom bounded only by the mountains.

When I first saw this central garden, the most extensive and regular of all the bee-pastures of the State, it seemed all one sheet of plant gold, hazy and vanishing in the distance, distinct as a new map along the foot-hills at my feet.

Descending the eastern slopes of the Coast Range through beds of gilias and lupines, and around many a breezy hillock and bush-crowned headland, I at length waded out into the midst of it. All the ground was covered, not with grass and green leaves, but with radiant corollas, about ankle-deep next the foot-hills, knee-deep or more five or six miles out. Here were bahia, madia, madaria, burrielia, chrysopsis, corethrogyne, grindelia, etc., growing in close social congregations of various shades of yellow, blending finely with the purples of clarkia, orthocarpus, and oenothera, whose delicate petals were drinking the vital sunbeams without giving back any sparkling glow.

Because so long a period of extreme drought succeeds the rainy season, most of the vegetation is composed of annuals, which spring up simultaneously, and bloom together at about the same height above the ground, the general surface being but slightly ruffled by the taller phacelias, pentstemons, and groups of *Salvia carduacea*, the king of the mints.

Sauntering in any direction, hundreds of these happy sun-plants brushed against my feet at every step, and closed over them as if I were wading in liquid gold. The air was sweet with fragrance, the larks sang their blessed songs, rising on the wing as I advanced, then sinking out of sight in the polleny sod, while myriads of wild bees stirred the lower air with their monotonous hum—monotonous, yet forever fresh and sweet as every-day sunshine. Hares and spermophiles showed themselves in considerable numbers in shallow places, and small bands of antelopes were almost constantly in sight, gazing curiously from some slight elevation, and then bounding swiftly away with unrivaled grace of motion. Yet I could discover no crushed flowers to mark their track, nor, indeed, any destructive action of any wild foot or tooth whatever.

The great yellow days circled by uncounted, while I drifted toward the north, observing the countless forms of life thronging about me, lying down almost anywhere on the approach of night. And what glorious botanical beds I had! Oftentimes on awaking I would find several new species leaning over me and looking me full in the face, so that my studies would begin before rising.

About the first of May I turned eastward, crossing the San Joaquin River between the mouths of the Tuolumne and Merced, and by the time I had reached the Sierra foot-hills most of the vegetation had gone to seed and become as dry as hay.

All the seasons of the great plain are warm or temperate, and bee-flowers are never wholly wanting; but the grand springtime—the annual resurrection—is governed by the rains, which usually set in about the middle of November or the beginning of December. Then the seeds, that for six months have lain on the ground dry and fresh as if they had been gathered into barns, at once unfold their treasured life. The general brown and purple of the ground, and the dead vegetation of the preceding year, give place to the green of mosses and liverworts and myriads of young leaves. Then one species after another comes into flower, gradually overspreading the green with yellow and purple, which lasts until May.

The "rainy season" is by no means a gloomy, soggy period of constant cloudiness and rain. Perhaps nowhere else in North America, perhaps in the world, are the months of December, January, February, and March so full of bland, plant-building sunshine. Referring to my notes of the winter and spring of 1868–69, every day of which I spent out of doors, on that section of the plain lying between the Tuolumne and Merced rivers, I find that the first rain of the season fell on December 18th. January had only six rainy days—that is, days on which rain fell; February three, March five, April three, and May three, completing the so-called rainy season, which was about an average one. The ordinary rain-storm of this region is seldom very cold or violent. The winds, which in settled weather come from the northwest, veer round into the opposite direction, the sky fills gradually and evenly with one general cloud, from which the rain falls steadily, often for days in succession, at a temperature of about 45° or 50°.

More than seventy-five per cent. of all the rain of this season came from the northwest, down the coast over southeastern Alaska, British Columbia, Washington, and Oregon, though the local winds of these circular storms blow from the southeast. One magnificent local storm from the northwest fell on March 21. A massive, round-browed cloud came swelling and thundering over the flowery plain in most imposing majesty, its bossy front burning white and purple in the full blaze of the sun, while warm rain poured from its ample fountains like a cataract, beating down flowers and bees, and flooding the dry watercourses as suddenly as those of Nevada are flooded by the so-called "cloudbursts." But in less than half an hour not a trace of the heavy, mountainlike cloud-structure was left in the sky, and the bees were on the wing, as if nothing more gratefully refreshing could have been sent them.

By the end of January four species of plants were in flower, and five or six mosses had already adjusted their hoods and were in the prime of life; but the flowers were not sufficiently numerous as yet to affect greatly the general green of the young leaves. Violets made their appearance in the first week of February, and toward the end of this month the warmer portions of the plain were already golden with myriads of the flowers of rayed compositae.

This was the full springtime. The sunshine grew warmer and richer, new plants bloomed every day; the air became more tuneful with humming wings, and sweeter with the fragrance of the opening flowers. Ants and ground squirrels were getting ready for their summer work, rubbing their benumbed limbs, and sunning themselves on the husk-piles before their doors, and spiders were busy mending their old webs, or weaving new ones.

In March, the vegetation was more than doubled in depth and color; claytonia, calandrinia, a large white gilia, and two nemophilas were in bloom, together with a host of yellow compositae, tall enough now to bend in the wind and show wavering ripples of shade.

In April, plant-life, as a whole, reached its greatest height, and the plain, over all its varied surface, was mantled with a close, furred plush of purple and golden corollas. By the end of this month, most of the species had ripened their seeds, but undecayed, still seemed to be in bloom from the numerous corolla-like involucres and whorls of chaffy scales of the compositae. In May, the bees found in flower only a few deep-set liliaceous plants and eriogonums.

June, July, August, and September is the season of rest and sleep,— a winter of dry heat,—followed in October by a second outburst of bloom at the very driest time of the year. Then, after the shrunken mass of leaves and stalks of the dead vegetation crinkle and turn to dust beneath the foot, as if it had been baked in an oven, *Hemizonia virgata*, a slender, unobtrusive little plant, from six inches to three feet high, suddenly makes its appearance in patches miles in extent, like a resurrection of the bloom in April. I have counted upward of 3000 flowers, five eighths of an inch in diameter, on a single plant. Both its leaves and stems are so slender as to be nearly invisible, at a distance of a few yards, amid so showy a multitude of flowers. The ray and disk flowers are both yellow, the stamens purple, and the texture of the rays is rich and velvety, like the petals of garden pansies. The prevailing wind turns all the heads round to the southeast, so that in facing northwestward we have the flowers looking us in the face. In my estimation, this little plant, the last born of the brilliant host of compositae that glorify the plain, is the most interesting of all. It remains in flower until November, uniting with two or three species of wiry eriogonums, which continue the floral chain around December to the spring flowers of January. Thus, although the main bloom and honey season is only about three months long, the floral circle, however thin around some of the hot, rainless months, is never completely broken.

How long the various species of wild bees have lived in this honey-garden, nobody knows; probably ever since the main body of the present flora gained possession of the land, toward the close of the glacial period. The first brown honey-bees brought to California are said to have arrived in San Francisco in March, 1853. A bee-keeper by the name of Shelton purchased a lot, consisting of twelve swarms, from some one at Aspinwall, who had brought them from New York. When landed at San Francisco, all the hives contained live bees, but they finally dwindled to one hive, which was taken to San José. The little immigrants flourished and multiplied in the bountiful pastures of the

Santa Clara Valley, sending off three swarms the first season. The owner was killed shortly afterward, and in settling up his estate, two of the swarms were sold at auction for $105 and $110 respectively. Other importations were made, from time to time, by way of the Isthmus, and, though great pains were taken to insure success, about one half usually died on the way. Four swarms were brought safely across the plains in 1859, the hives being placed in the rear end of a wagon, which was stopped in the afternoon to allow the bees to fly and feed in the floweriest places that were within reach until dark, when the hives were closed.

In 1855, two years after the time of the first arrivals from New York, a single swarm was brought over from San José, and let fly in the Great Central Plain. Bee-culture, however, has never gained much attention here, notwithstanding the extraordinary abundance of honey-bloom, and the high price of honey during the early years. A few hives are found here and there among settlers who chanced to have learned something about the business before coming to the State. But sheep, cattle, grain, and fruit raising are the chief industries, as they require less skill and care, while the profits thus far have been greater. In 1856 honey sold here at from one and a half to two dollars per pound. Twelve years later the price had fallen to twelve and a half cents. In 1868 I sat down to dinner with a band of ravenous sheep-shearers at a ranch on the San Joaquin, where fifteen or twenty hives were kept, and our host advised us not to spare the large pan of honey he had placed on the table, as it was the cheapest article he had to offer. In all my walks, however, I have never come upon a regular bee-ranch in the Central Valley like those so common and so skilfully managed in the southern counties of the State. The few pounds of honey and wax produced are consumed at home, and are scarcely taken into account among the coarser products of the farm. The swarms that escape from their careless owners have a weary, perplexing time of it in seeking suitable homes. Most of them make their way to the foot-hills of the mountains, or to the trees that line the banks of the rivers, where some hollow log or trunk may be found. A friend of mine, while out hunting on the San Joaquin, came upon an old coon trap, hidden among some tall grass, near the edge of the river, upon which he sat down to rest. Shortly afterward his attention was attracted to a crowd of angry bees that were flying excitedly about his head, when he discovered that he was sitting upon their hive, which was found to contain more than 200 pounds of honey. Out in the broad, swampy delta of the Sacramento and San Joaquin rivers, the little wanderers have been known to build their combs in a bunch of rushes, or stiff, wiry grass, only slightly protected from the weather, and in danger every spring of being carried away by floods. They have the advantage, however, of a vast extent of fresh pasture, accessible only to themselves.

The present condition of the Grand Central Garden is very different from that we have sketched. About twenty years ago, when the gold placers had been pretty thoroughly exhausted, the attention of fortune-seekers—not home-seekers—was, in great part, turned away from the mines to the fertile plains, and many began experiments in a kind of restless, wild agriculture. A load of lumber would be hauled to some spot on the free wilderness, where water could be easily found, and a rude box-cabin built. Then a gang-plow was procured, and a dozen mustang ponies, worth ten or fifteen dollars apiece, and with these hundreds of acres were stirred as easily as if the land had been under cultivation for years, tough, perennial roots being almost wholly absent. Thus a ranch was established, and from these bare wooden huts, as centers of desolation, the wild flora vanished in ever-widening circles. But the arch destroyers are the shepherds, with their flocks of hoofed locusts, sweeping over the ground like a fire, and trampling down every rod that escapes the plow as completely as if the whole plain were a cottage garden-plot without a fence. But notwithstanding these destroyers, a thousand swarms of bees may be pastured here for every one now gathering honey. The greater portion is still covered every season with a repressed growth of bee-flowers, for most of the species are annuals, and many of them are not relished by sheep or cattle, while the rapidity of their growth enables them to develop and mature their seeds before any foot has time to crush them. The ground is, therefore, kept sweet, and the race is perpetuated, though only as a suggestive shadow of the magnificence of its wildness.

The time will undoubtedly come when the entire area of this noble valley will be tilled like a garden, when the fertilizing waters of the mountains, now flowing to the sea, will be distributed to every acre, giving rise to prosperous towns, wealth, arts, etc. Then, I suppose, there will be few left, even among botanists, to deplore the vanished primeval flora. In the mean time, the pure waste going on—the wanton destruction of the innocents—is a sad sight to see, and the sun may well be pitied in being compelled to look on.

The bee-pastures of the Coast Ranges last longer and are more varied than those of the great plain, on account of differences of soil and climate, moisture, and shade, etc. Some of the mountains are upward of 4000 feet in height, and small streams, springs, oozy bogs, etc., occur in great abundance and variety in the wooded regions, while open parks, flooded with sunshine, and hill-girt valleys lying at different elevations, each with its own peculiar climate and exposure, possess the required conditions for the development of species and families of plants widely varied.

Next the plain there is, first, a series of smooth hills, planted with a rich and showy vegetation that differs but little from that of the plain itself—as if the edge of the plain had been lifted and bent into flowing folds, with all its flowers in place, only toned down a little as to their luxuriance, and a few new species introduced, such as the hill lupines, mints, and gilias. The colors show finely when thus held to view on the slopes; patches of red, purple, blue, yellow, and white, blending around the edges, the whole appearing at a little distance like a map colored in sections.

Above this lies the park and chaparral region, with oaks, mostly evergreen, planted wide apart, and blooming shrubs from three to ten feet high; manzanita and ceanothus of several species, mixed with rhamnus, cercis, pickeringia, cherry, amelanchier, and adenostoma, in shaggy, interlocking thickets, and many species of hosackia, clover, monardella, castilleia, etc., in the openings.

The main ranges send out spurs somewhat parallel to their axes, inclosing level valleys, many of them quite extensive, and containing a great profusion of sun-loving bee-flowers in their wild state; but these are, in great part, already lost to the bees by cultivation.

Nearer the coast are the giant forests of the redwoods, extending from near the Oregon line to Santa Cruz. Beneath the cool, deep shade of these majestic trees the ground is occupied by ferns, chiefly woodwardia and aspidiums, with only a few flowering plants—oxalis, trientalis, erythronium, fritillaria, smilax, and other shade-lovers. But all along the redwood belt there are sunny openings on hill-slopes looking to the south, where the giant trees stand back, and give the ground to the small sunflowers and the bees. Around the lofty redwood walls of these little bee-acres there is usually a fringe of Chestnut Oak, Laurel, and Madroño, the last of which is a surpassingly beautiful tree, and a great favorite with the bees. The trunks of the largest specimens are seven or eight feet thick, and about fifty feet high; the bark red and chocolate colored, the leaves plain, large, and glossy, like those of *Magnolia grandiflora*, while the flowers are yellowish-white, and urn-shaped, in well-proportioned panicles, from five to ten inches long. When in full bloom, a single tree seems to be visited at times by a whole hive of bees at once, and the deep hum of such a multitude makes the listener guess that more than the ordinary work of honey-winning must be going on.

How perfectly enchanting and care-obliterating are these with-drawn gardens of the woods—long vistas opening to the sea—sunshine sifting and pouring upon the flowery ground in a tremulous, shifting mosaic, as the light-ways in the leafy wall open and close with the swaying breeze—shining leaves and flowers, birds and bees, mingling together in springtime harmony, and soothing fragrance exhaling from a thousand thousand fountains! In these balmy, dissolving days, when the deep heart-beats of Nature are felt thrilling rocks and trees and everything alike, common business and friends are happily forgotten, and even the natural honey-work of bees, and the care of birds for their young, and mothers for their children, seem slightly out of place.

To the northward, in Humboldt and the adjacent counties, whole hillsides are covered with rhododendron, making a glorious melody of bee-bloom in the spring. And the Western azalea, hardly less flowery, grows in massy thickets three to eight feet high around the edges of groves and woods as far south as San Luis Obispo, usually accom-panied by manzanita; while the valleys, with their varying moisture and shade, yield a rich variety of the smaller honey-flowers, such as mentha, lycopus, micromeria, audibertia, trichostema, and other mints; with vaccinium, wild strawberry, geranium, calais, and goldenrod; and in the cool glens along the stream-banks, where the shade of trees is not too deep, spiraea, dog-wood, heteromeles, and calycanthus, and many species of rubus form interlacing tangles, some portion of which continues in bloom for months.

Though the coast region was the first to be invaded and settled by white men, it has suffered less from a bee point of view than either of the other main divisions, chiefly, no doubt, because of the unevenness of the surface, and because it is owned and protected instead of lying exposed to the flocks of the wandering "sheepmen." These remarks apply more particularly to the north half of the coast. Farther south there is less moisture, less forest shade, and the honey flora is less varied.

The Sierra region is the largest of the three main divisions of the bee-lands of the State, and the most regularly varied in its subdivisions, owing to their gradual rise from the level of the Central Plain to the alpine summits. The foot-hill region is about as dry and sunful, from the end of May until the setting in of the winter rains, as the plain. There are no shady forests, no damp glens, at all like those lying at the same elevations in the Coast Mountains. The social compositae of the plain, with a few added species, form the bulk of the herbaceous portion of the vegetation up to a height of 1500 feet or more, shaded lightly here and there with oaks and Sabine Pines, and interrupted by patches of ceanothus and buckeye. Above this, and just below the forest region, there is a dark, heath-like belt of chaparral, composed almost exclusively of *Adenostoma fasciculata*, a bush belonging to the rose family, from five to eight feet high, with small, round leaves in fascicles, and bearing a multitude of small white flowers in panicles on the ends of the upper branches. Where it occurs at all, it usually covers all the ground with a close, impenetrable growth, scarcely broken for miles.

Up through the forest region, to a height of about 9000 feet above sea-level, there are ragged patches of manzanita, and five or six species of ceanothus, called deer-brush or California lilac. These are the most important of all the honey-bearing bushes of the Sierra. *Chamoebatia foliolosa*, a little shrub about a foot high, with flowers like the strawberry, makes handsome carpets beneath the pines, and seems to be a favorite with the bees; while pines themselves furnish unlimited quantities of pollen and honey-dew. The product of a single tree, ripening its pollen at the right time of year, would be sufficient for the wants of a whole hive. Along the streams there is a rich growth of lilies, larkspurs, pedicularis, castilleias, and clover. The alpine region contains the flowery glacier meadows, and countless small gardens in all sorts of places full of potentilla of several species, spraguea, ivesia, epilobium, and goldenrod, with beds of bryanthus and the charming cassiope covered with sweet bells. Even the tops of the mountains are blessed with flowers,—dwarf phlox, polemonium, ribes, hulsea, etc. I have seen wild bees and butterflies feeding at a height of 13,000 feet above the sea. Many, however, that go up these dangerous heights never come down again. Some, undoubtedly, perish in storms, and I have found thousands lying dead or benumbed on the surface of the glaciers, to which they had perhaps been attracted by the white glare, taking them for beds of bloom.

From swarms that escaped their owners in the lowlands, the honey-bee is now generally distributed throughout the whole length of the Sierra, up to an elevation of 8000 feet above sea-level. At this height they flourish without care, though the snow every winter is deep. Even higher than this several bee-trees have been cut which contained over 200 pounds of honey.

The destructive action of sheep has not been so general on the mountain pastures as on those of the great plain, but in many places it has been more complete, owing to the more friable character of the soil, and its sloping position. The slant digging and down-raking action of hoofs on the steeper slopes of moraines has uprooted and buried many of the tender plants from year to year, without allowing them time to mature their seeds. The shrubs, too, are badly bitten, especially the various species of ceanothus. Fortunately, neither sheep nor cattle care to feed on the manzanita, spiraea, or adenostoma; and these fine honey-bushes are too stiff and tall, or grow in places too rough and inaccessible, to be trodden under foot. Also the cañon walls and gorges, which form so considerable a part of the area of the range, while inaccessible to domestic sheep, are well fringed with honey-shrubs, and contain thousands of lovely bee-gardens, lying hid in narrow side-cañons and recesses fenced with avalanche taluses, and on the top of flat, projecting headlands, where only bees would think to look for them.

But, on the other hand, a great portion of the woody plants that escape the feet and teeth of the sheep are destroyed by the shepherds by means of running fires, which are set everywhere during the dry autumn for the purpose of burning off the old fallen trunks and under-brush, with a view to improving the pastures, and making more open ways for the flocks. These destructive sheep-fires sweep through nearly the entire forest belt of the range, from one extremity to the other, consuming not only the underbrush, but the young trees and seedlings on which the permanence of the forests depends; thus setting in motion a long train of evils which will certainly reach far beyond bees and bee-keepers.

The plow has not yet invaded the forest region to any appreciable extent, neither has it accomplished much in the foot-hills. Thousands of bee-ranches might be established along the margin of the plain, and up to a height of 4000 feet, wherever water could be obtained. The climate at this elevation admits of the making of permanent homes, and by moving the hives to higher pastures as the lower pass out of bloom, the annual yield of honey would be nearly doubled. The foot-hill pastures, as we have seen, fail about the end of May, those of the chaparral belt and lower forests are in full bloom in June, those of the upper and alpine region in July, August, and September. In Scotland, after the best of the Lowland bloom is past, the bees are carried in carts to the Highlands, and set free on the heather hills. In France, too, and in Poland, they are carried from pasture to pasture among orchards and fields in the same way, and along the rivers in barges to collect the honey of the delightful vegetation of the banks. In Egypt they are taken far up the Nile, and floated slowly home again, gathering the honey-harvest of the various fields on the way, timing their movements in accord with the seasons. Were similar methods pursued in California the productive season would last nearly all the year.

The average elevation of the north half of the Sierra is, as we have seen, considerably less than that of the south half, and small streams, with the bank and meadow gardens dependent upon them, are less abundant. Around the head waters of the Yuba, Feather, and Pitt rivers, the extensive tablelands of lava are sparsely planted with pines, through which the sunshine reaches the ground with little interruption. Here flourishes a scattered, tufted growth of golden applopappus, linosyris, bahia, wyetheia, arnica, artemisia, and similar plants; with manzanita, cherry, plum, and thorn in ragged patches on the cooler hill-slopes. At the extremities of the Great Central Plain, the Sierra and Coast Ranges curve around and lock together in a labyrinth of mountains and valleys, throughout which their floras are mingled, making at the north, with its temperate climate and copious rainfall, a perfect paradise for bees, though, strange to say, scarcely a single regular bee-ranch has yet been established in it.

Of all the upper flower fields of the Sierra, Shasta is the most honeyful, and may yet surpass in fame the celebrated honey hills of Hybla and hearthy Hymettus. Regarding this noble mountain from a bee point of view, encircled by its many climates, and sweeping aloft from the torrid plain into the frosty azure, we find the first 5000 feet from the summit generally snow-clad, and therefore about as honeyless as the sea. The base of this arctic region is girdled by a belt of crumbling lava measuring about 1000 feet in vertical breadth, and is mostly free from snow in summer. Beautiful lichens enliven the faces of the cliffs with their bright colors, and in some of the warmer nooks there are a few tufts of alpine daisies, wall-flowers and pentstemons; but, notwithstanding these bloom freely in the late summer, the zone as a whole is almost as honeyless as the icy summit, and its lower edge may be taken as the honey-line. Immediately below this comes the forest zone, covered with a rich growth of conifers, chiefly Silver Firs, rich in pollen and honey-dew, and diversified with countless garden openings, many of them less than a hundred yards across. Next, in orderly succession, comes the great bee zone. Its area far surpasses that of the icy summit and both the other zones combined, for it goes sweeping majestically around the entire mountain, with a breadth of six or seven miles and a circumference of nearly a hundred miles.

Shasta, as we have already seen, is a fire-mountain created by a succession of eruptions of ashes and molten lava, which, flowing over the lips of its several craters, grew outward and upward like the trunk of a knotty exogenous tree. Then followed a strange contrast. The glacial winter came on, loading the cooling mountain with ice, which flowed slowly outward in every direction, radiating from the summit in the form of one vast conical glacier—a down-crawling mantle of ice upon a fountain of smoldering fire, crushing and grinding for centuries its brown, flinty lavas with incessant activity, and thus degrading and remodeling the entire mountain. When, at length, the glacial period began to draw near its close, the ice-mantle was gradually melted off around the bottom, and, in receding and breaking into its present fragmentary condition, irregular rings and heaps of moraine matter were stored upon its flanks. The glacial erosion of most of the Shasta lavas produces detritus, composed of rough, sub-angular boulders of moderate size and of porous gravel and sand, which yields freely to the transporting power of running water. Magnificent floods from the ample fountains of ice and snow working with sublime energy upon this prepared glacial detritus, sorted it out and carried down immense quantities from the higher slopes, and reformed it in smooth, delta-like beds around the base; and it is these flood-beds joined together that now form the main honey-zone of the old volcano.

Thus, by forces seemingly antagonistic and destructive, has Mother Nature accomplished her beneficent designs—now a flood of fire, now a flood of ice, now a flood of water; and at length an outburst of organic life, a milky way of snowy petals and wings, girdling the rugged mountain like a cloud, as if the vivifying sunbeams beating against its sides had broken into a foam of plant-bloom and bees, as sea-waves break and bloom on a rock shore.

In this flowery wilderness the bees rove and revel, rejoicing in the bounty of the sun, clambering eagerly through bramble and huckle-bloom, ringing the myriad bells of the manzanita, now humming aloft among polleny willows and firs, now down on the ashy ground among gilias and buttercups, and anon plunging deep into snowy banks of cherry and buckthorn. They consider the lilies and roll into them, and, like lilies, they toil not, for they are impelled by sun-power, as water-wheels by water-power; and when the one has plenty of high-pressure water, the other plenty of sunshine, they hum and quiver alike. Saunter-ing in the Shasta bee-lands in the sun-days of summer, one may readily infer the time of day from the comparative energy of bee-movements alone—drowsy and moderate in the cool of the morning, increasing in energy with the ascending sun, and, at high noon, thrilling and quiver-ing in wild ecstasy, then gradually declining again to the stillness of night. In my excursions among the glaciers I occasionally meet bees that are hungry, like mountaineers who venture too far and remain too long above the bread-line; then they droop and wither like autumn leaves. The Shasta bees are perhaps better fed than any others in the Sierra. Their field-work is one perpetual feast; but, however exhilarating the sunshine or bountiful the supply of flowers, they are always dainty feeders. Humming-moths and humming-birds seldom set foot upon a flower, but poise on the wing in front of it, and reach forward as if they were sucking through straws. But bees, though as dainty as they, hug their favorite flowers with profound cordiality, and push their blunt, polleny faces against them, like babies on their mother's bosom. And fondly, too, with eternal love, does Mother Nature clasp her small bee-babies, and suckle them, multitudes at once, on her warm Shasta breast.

Besides the common honey-bee there are many other species here—fine mossy, burly fellows, who were nourished on the mountains thou-sands of sunny seasons before the advent of the domestic species. Among these are the bumblebees, mason-bees, carpenter-bees, and leaf-cutters. Butterflies, too, and moths of every size and pattern; some broad-winged like bats, flapping slowly, and sailing in easy curves; others like small, flying violets, shaking about loosely in short, crooked flights close to the flowers, feasting luxuriously night and day. Great numbers of deer also delight to dwell in the brushy portions of the bee-pastures.

Bears, too, roam the sweet wilderness, their blunt, shaggy forms harmonizing well with the trees and tangled bushes, and with the bees, also, notwithstanding the disparity in size. They are fond of all good things, and enjoy them to the utmost, with but little troublesome discrimination—flowers and leaves as well as berries, and the bees themselves as well as their honey. Though the California bears have as yet had but little experience with honey-bees, they often succeed in reaching their bountiful stores, and it seems doubtful whether bees themselves enjoy honey with so great a relish. By means of their powerful teeth and claws they can gnaw and tear open almost any hive conveniently accessible. Most honey-bees, however, in search of a home are wise enough to make choice of a hollow in a living tree, a considerable distance above the ground, when such places are to be had; then they are pretty secure, for though the smaller black and brown bears climb well, they are unable to break into strong hives while compelled to exert themselves to keep from falling, and at the same time to endure the stings of the fighting bees without having their paws free to rub them off. But woe to the black bumblebees discovered in their mossy nests in the ground! With a few strokes of their huge paws the bears uncover the entire establishment, and, before time is given for a general buzz, bees old and young, larvae, honey, stings, nest, and all are taken in one ravishing mouthful.

Not the least influential of the agents concerned in the superior sweetness of the Shasta flora are its storms—storms I mean that are strictly local, bred and born on the mountain. The magical rapidity with which they are grown on the mountain-top, and bestow their charity in rain and snow, never fails to astonish the inexperienced lowlander. Often in calm, glowing days, while the bees are still on the wing, a storm-cloud may be seen far above in the pure ether, swelling its pearl bosses, and growing silently, like a plant. Presently a clear, ringing discharge of thunder is heard, followed by a rush of wind that comes sounding over the bending woods like the roar of the ocean, mingling raindrops, snow-flowers, honey-flowers, and bees in wild storm harmony.

Still more impressive are the warm, reviving days of spring in the mountain pastures. The blood of the plants throbbing beneath the life-giving sunshine seems to be heard and felt. Plant growth goes on before our eyes, and every tree in the woods, and every bush and flower is seen as a hive of restless industry. The deeps of the sky are mottled with singing wings of every tone and color; clouds of brilliant chry-sididae dancing and swirling in exquisite rhythm, golden-barred ves-pidae, dragon-flies, butterflies, grating cicadas, and jolly, rattling grass-hoppers, fairly enameling the light.

On bright, crisp mornings a striking optical effect may frequently be observed from the shadows of the higher mountains while the sun-beams are pouring past overhead. Then every insect, no matter what may be its own proper color, burns white in the light. Gauzy-winged hymenoptera, moths, jet-black beetles, all are transfigured alike in pure, spiritual white, like snowflakes.

In Southern California, where bee-culture has had so much skilful attention of late years, the pasturage is not more abundant, or more advantageously varied as to the number of its honey-plants and their distribution over mountain and plain, than that of many other portions of the State where the industrial currents flow in other channels. The famous White Sage (*Audibertia*), belonging to the mint family, flour-ishes here in all its glory, blooming in May, and yielding great quanti-ties of clear, pale honey, which is greatly prized in every market it has yet reached. This species grows chiefly in the valleys and low hills. The Black Sage on the mountains is part of a dense, thorny chaparral, which is composed chiefly of adenostoma, ceanothus, manzanita, and cherry—not differing greatly from that of the southern portion of the Sierra, but more dense and continuous, and taller, and remaining longer in bloom. Stream-side gardens, so charming a feature of both the Sierra and Coast Mountains, are less numerous in Southern California, but they are exceedingly rich in honey-flowers, wherever found,—melilotus, columbine, collinsia, verbena, zauschneria, wild rose, honeysuckle, philadelphus, and lilies rising from the warm, moist dells in a very storm of exuberance. Wild buckwheat of many species is developed in abun-dance over the dry, sandy valleys and lower slopes of the mountains, toward the end of summer, and is, at this time, the main dependence of the bees, reinforced here and there by orange groves, alfalfa fields, and small home gardens.

The main honey months, in ordinary seasons, are April, June, July, and August; while the other months are usually flowery enough to yield sufficient for the bees.

According to Mr. J. T. Gordon, President of the Los Angeles County Bee-keepers' Association, the first bees introduced into the county were a single hive, which cost $150 in San Francisco, and arrived in September, 1854.[1] In April, of the following year, this hive sent out two swarms, which were sold for $100 each. From this small beginning the bees gradually multiplied to about 3000 swarms in the year 1873. In 1876 it was estimated that there were between 15,000 and 20,000 hives in the county, producing an annual yield of about 100 pounds to the hive—in some exceptional cases, a much greater yield.

In San Diego County, at the beginning of the season of 1878, there were about 24,000 hives, and the shipments from the one port of San Diego for the same year, from July 17 to November 10, were 1071 barrels, 15,544 cases, and nearly 90 tons. The largest bee-ranches have about a thousand hives, and are carefully and skilfully managed, every scientific appliance of merit being brought into use. There are few bee-keepers, however, who own half as many as this, or who give their undivided attention to the business. Orange culture, at present, is heavily overshadowing every other business.

A good many of the so-called bee-ranches of Los Angeles and San Diego counties are still of the rudest pioneer kind imaginable. A man unsuccessful in everything else hears the interesting story of the profits and comforts of bee-keeping, and concludes to try it; he buys a few colonies, or gets them from some overstocked ranch on shares, takes them back to the foot of some cañon, where the pasturage is fresh, squats on the land, with, or without, the permission of the owner, sets up his hives, makes a box-cabin for himself, scarcely bigger than a bee-hive, and awaits his fortune.

Bees suffer sadly from famine during the dry years which occasionally occur in the southern and middle portions of the State. If the rainfall amounts only to three or four inches, instead of from twelve to twenty, as in ordinary seasons, then sheep and cattle die in thousands, and so do these small, winged cattle, unless they are carefully fed, or removed to other pastures. The year 1877 will long be remembered as exceptionally rainless and distressing. Scarcely a flower bloomed on the dry valleys away from the stream-sides, and not a single grain-field depending upon rain was reaped. The seed only sprouted, came up a little way, and withered. Horses, cattle, and sheep grew thinner day by day, nibbling at bushes and weeds, along the shallowing edges of

[1] Fifteen hives of Italian bees were introduced into Los Angeles County in 1855, and in 1876 they had increased to 500. The marked superiority claimed for them over the common species is now attracting considerable attention.

streams, many of which were dried up altogether, for the first time since the settlement of the country.

In the course of a trip I made during the summer of that year through Monterey, San Luis Obispo, Santa Barbara, Ventura, and Los Angeles counties, the deplorable effects of the drought were everywhere visible—leafless fields, dead and dying cattle, dead bees, and half-dead people with dusty, doleful faces. Even the birds and squirrels were in distress, though their suffering was less painfully apparent than that of the poor cattle. These were falling one by one in slow, sure starvation along the banks of the hot, sluggish streams, while thousands of buzzards correspondingly fat were sailing above them, or standing gorged on the ground beneath the trees, waiting with easy faith for fresh carcasses. The quails, prudently considering the hard times, abandoned all thought of pairing. They were too poor to marry, and so continued in flocks all through the year without attempting to rear young. The ground-squirrels, though an exceptionally industrious and enterprising race, as every farmer knows, were hard pushed for a living; not a fresh leaf or seed was to be found save in the trees, whose bossy masses of dark green foliage presented a striking contrast to the ashen baldness of the ground beneath them. The squirrels, leaving their accustomed feeding-grounds, betook themselves to the leafy oaks to gnaw out the acorn stores of the provident woodpeckers, but the latter kept up a vigilant watch upon their movements. I noticed four woodpeckers in league against one squirrel, driving the poor fellow out of an oak that they claimed. He dodged round the knotty trunk from side to side, as nimbly as he could in his famished condition, only to find a sharp bill everywhere. But the fate of the bees that year seemed the saddest of all. In different portions of Los Angeles and San Diego counties, from one half to three fourths of them died of sheer starvation. Not less than 18,000 colonies perished in these two counties alone, while in the adjacent counties the death-rate was hardly less.

Even the colonies nearest to the mountains suffered this year, for the smaller vegetation on the foot-hills was affected by the drought almost as severely as that of the valleys and plains, and even the hardy, deep-rooted chaparral, the surest dependence of the bees, bloomed sparingly, while much of it was beyond reach. Every swarm could have been saved, however, by promptly supplying them with food when their own stores began to fail, and before they became enfeebled and discouraged; or by cutting roads back into the mountains, and taking them into the heart of the flowery chaparral. The Santa Lucia, San Rafael, San Gabriel, San Jacinto, and San Bernardino ranges are almost untouched as yet save by the wild bees. Some idea of their resources, and of the advantages and disadvantages they offer to bee-keepers, may be formed from an excursion that I made into the San Gabriel Range about the beginning of August of "the dry year." This range, containing most of the characteristic features of the other ranges just mentioned, overlooks the Los Angeles vineyards and orange groves from the north, and is more rigidly inaccessible in the ordinary meaning of the word than any other that I ever attempted to penetrate. The slopes are exceptionally steep and insecure to the foot, and they are covered with thorny bushes from five to ten feet high. With the exception of little spots not visible in general views, the entire surface is covered with them, massed in close hedge growth, sweeping gracefully down into every gorge and hollow, and swelling over every ridge and summit in shaggy, ungovernable exuberance, offering more honey to the acre for half the year than the most crowded cloverfield. But when beheld from the open San Gabriel Valley, beaten with dry sunshine, all that was seen of the range seemed to wear a forbidding aspect. From base to summit all seemed gray, barren, silent, its glorious chaparral appearing like dry moss creeping over its dull, wrinkled ridges and hollows.

Setting out from Pasadena, I reached the foot of the range about sundown; and being weary and heated with my walk across the shadeless valley, concluded to camp for the night. After resting a few moments, I began to look about among the flood-boulders of Eaton Creek for a camp-ground, when I came upon a strange, dark-looking man who had been chopping cord-wood. He seemed surprised at seeing me, so I sat down with him on the live-oak log he had been cutting, and made haste to give a reason for my appearance in his solitude, explaining that I was anxious to find out something about the mountains, and meant to make my way to Eaton Creek next morning. Then he kindly invited me to camp with him, and led me to his little cabin, situated at the foot of the mountains, where a small spring oozes out of a bank overgrown with wild-rose bushes. After supper, when the daylight was gone, he explained that he was out of candles; so we sat in the dark, while he gave me a sketch of his life in a mixture of Spanish and English. He was born in Mexico, his father Irish, his mother Spanish. He had been a miner, rancher, prospector, hunter, etc., rambling always, and wearing his life away in mere waste; but now he was going to settle down. His past life, he said, was of "no account," but the future was promising. He was going to "make money and marry a Spanish woman." People mine here for water as for gold. He had been running a tunnel into a spur of the mountain back of his cabin. "My prospect is good," he said, "and if I chance to strike a good, strong flow, I'll soon be worth $5000 or $10,000. For that flat out there," referring to a small, irregular patch of bouldery detritus, two or three acres in size, that had been deposited by Eaton Creek during some flood season,—"that flat is large enough for a nice orange-grove, and the bank behind the cabin will do for a vineyard, and after watering my own trees and vines I will have some water left to sell to my neighbors below me, down the valley. And then," he continued, "I can keep bees, and make money that way, too, for the mountains above here are just full of honey in the summertime, and one of my neighbors down here says that he will let me have a whole lot of hives, on shares, to start with. You see I've a good thing; I'm all right now." All this prospective affluence in the sunken, boulder-choked flood-bed of a mountain-stream! Leaving the bees out of the count, most fortune-seekers would as soon think of settling on the summit of Mount Shasta. Next morning, wishing my hopeful entertainer good luck, I set out on my shaggy excursion.

About half an hour's walk above the cabin, I came to "The Fall," famous throughout the valley settlements as the finest yet discovered in the San Gabriel Mountains. It is a charming little thing, with a low, sweet voice, singing like a bird, as it pours from a notch in a short ledge, some thirty-five or forty feet into a round mirror-pool. The face of the cliff back of it, and on both sides, is smoothly covered and embossed with mosses, against which the white water shines out in showy relief, like a silver instrument in a velvet case. Hither come the San Gabriel lads and lassies, to gather ferns and dabble away their hot holidays in the cool water, glad to escape from their commonplace palm-gardens and orange-groves. The delicate maidenhair grows on fissured rocks within reach of the spray, while broad-leaved maples and sycamores cast soft, mellow shade over a rich profusion of bee-flowers, growing among boulders in front of the pool—the fall, the flowers, the bees, the ferny rocks, and leafy shade forming a charming little poem of wildness, the last of a series extending down the flowery slopes of Mount San Antonio through the rugged, foam-beaten bosses of the main Eaton Cañon.

From the base of the fall I followed the ridge that forms the western rim of the Eaton basin to the summit of one of the principle peaks, which is about 5000 feet above sea-level. Then, turning eastward, I crossed the middle of the basin, forcing a way over its many subordinate ridges and across its eastern rim, having to contend almost everywhere with the floweriest and most impenetrable growth of honeybushes I had ever encountered since first my mountaineering began. Most of the Shasta chaparral is leafy nearly to the ground; here the main stems are naked for three or four feet, and interspiked with dead twigs, forming a stiff *chevaux de frise* through which even the bears make their way with difficulty. I was compelled to creep for miles on all fours, and in following the bear-trails often found tufts of hair on the bushes where they had forced themselves through.

For 100 feet or so above the fall the ascent was made possible only by tough cushions of club-moss that clung to the rock. Above this the ridge weathers away to a thin knife-blade for a few hundred yards, and thence to the summit of the range it carries a bristly mane of chaparral. Here and there small openings occur on rocky places, commanding fine views across the cultivated valley to the ocean. These I found by the tracks were favorite outlooks and resting-places for the wild animals— bears, wolves, foxes, wildcats, etc.—which abound here, and would have to be taken into account in the establishment of bee-ranches. In the deepest thickets I found wood-rat villages—groups of huts four to six feet high, built of sticks and leaves in rough, tapering piles, like musk-rat cabins. I noticed a good many bees, too, most of them wild. The tame honey-bees seemed languid and wing-weary, as if they had come all the way up from the flowerless valley.

After reaching the summit I had time to make only a hasty survey of the basin, now glowing in the sunset gold, before hastening down into one of the tributary cañons in search of water. Emerging from a particularly tedious breadth of chaparral, I found myself free and erect in a beautiful park-like grove of Mountain Live Oak, where the ground was planted with aspidiums and brier-roses, while the glossy foliage made a close canopy overhead, leaving the gray dividing trunks bare to show the beauty of their interlacing arches. The bottom of the cañon was dry where I first reached it, but a bunch of scarlet mimulus indicated water at no great distance, and I soon discovered about a bucketful in a hollow of the rock. This, however, was full of dead bees, wasps, beetles, and leaves, well steeped and simmered, and would, therefore, require boiling and filtering through fresh charcoal before it could be made available. Tracing the dry channel about a mile farther down to its junction with a larger tributary cañon, I at length discovered a lot of boulder pools, clear as crystal, brimming full, and linked together by glistening streamlets just strong enough to sing audibly. Flowers in full bloom adorned their margins, lilies ten feet high, larkspur, columbines, and luxuriant ferns, leaning and overarching in lavish abundance, while a noble old Live Oak spread its rugged arms over all. Here I camped, making my bed on smooth cobblestones.

Next day, in the channel of a tributary that heads on Mount San Antonio, I passed about fifteen or twenty gardens like the one in which I slept—lilies in every one of them, in the full pomp of bloom. My third camp was made near the middle of the general basin, at the head of a long system of cascades from ten to 200 feet high, one following the other in close succession down a rocky, inaccessible cañon, making a total descent of nearly 1700 feet. Above the cascades the main stream passes through a series of open, sunny levels, the largest of which are about an acre in size, where the wild bees and their companions were feasting on a showy growth of zauschneria, painted cups, and monardella; and gray squirrels were busy harvesting the burs of the Douglas Spruce, the only conifer I met in the basin.

The eastern slopes of the basin are in every way similar to those we have described, and the same may be said of other portions of the range. From the highest summit, far as the eye could reach, the landscape was one vast bee-pasture, a rolling wilderness of honey-bloom, scarcely broken by bits of forest or the rocky outcrops of hilltops and ridges.

Behind the San Bernardino Range lies the wild "sage-brush country," bounded on the east by the Colorado River, and extending in a general northerly direction to Nevada and along the eastern base of the Sierra beyond Mono Lake.

The greater portion of this immense region, including Owen's Valley, Death Valley, and the Sink of the Mohave, the area of which is nearly one fifth that of the entire State, is usually regarded as a desert, not because of any lack in the soil, but for want of rain, and rivers available for irrigation. Very little of it, however, is desert in the eyes of a bee.

Looking now over all the available pastures of California, it appears that the business of bee-keeping is still in its infancy. Even in the more enterprising of the southern counties, where so vigorous a beginning has been made, less than a tenth of their honey resources have as yet been developed; while in the Great Plain, the Coast Ranges, the Sierra Nevada, and the northern region about Mount Shasta, the business can hardly be said to exist at all. What the limits of its developments in the future may be, with the advantages of cheaper transportation and the invention of better methods in general, it is not easy to guess. Nor, on the other hand, are we able to measure the influence on bee interests likely to follow the destruction of the forests, now rapidly falling before fire and the ax. As to the sheep evil, that can hardly become greater than it is at the present day. In short, notwithstanding the wide-spread deterioration and destruction of every kind already effected, California, with her incomparable climate and flora, is still, as far as I know, the best of all the bee-lands of the world.

WATCHERS
AT THE POND *

FRANKLIN RUSSELL

Franklin Russell is a free-lance writer in New Zealand and
Australia who is also a born biologist. He was brought up on
a farm in New Zealand and was at one with nature as a boy.
Russell spent one year at a pond studying the life and death
struggle between animals which goes on almost unnoticed by
the uninitiated observer. It is, he says,

> ... a haunting echo of the past ... preserved ...
> every detail of the spring awakening; the reaching for
> space, light, expression; the langorous heat of summer
> days; the slow waning metabolism of the last season;
> and the long sleep away from the sun. In every speck
> of living matter, there was this memory of the inde-
> structible life force of earth.

THE WATCHERS IN THE SNOW

At the beginning of that year, a red-tailed hawk paused high over the frozen pond.

He wheeled slowly and the earth revolved under him. He saw the wing-shaped scar of the pond set in a still and fathomless variegation of bare trees, ravines, dark clumps of pines, and wandering white outlines of frozen streams. He floated under wide wings that soughed in the north wind, a sound that was rough-edged as he turned across wind, then smooth and keen as he cut down into its persistent force. His yellow legs bunched tightly under his belly, and his orange nostrils glowed in the cold sun. He knew a small part of the secret of the pond.

He saw the indistinct outline of a snowshoe hare standing at the edge of the pond. He heard the call of a dog fox and saw the bitch running through some trees. He saw the bare oak, north of the pond, where he had been born in a previous season. He saw a great marsh, south of the pond, over which his sister and parents had headed to the ocean during the time of migration.

His extraordinary eyesight enabled him to see the pond in fine detail and to find movement in the stillness and meaning in flickering specks of color in thickets where tiny golden-crowned kinglets and chickadees darted between trees. He heard the tinkling, chippering cries of sparrows, rising like broken crystals. He saw the trunk circlers—woodpeckers, creepers, nuthatches—and heard their peeking calls and the precise ticks of their beaks driving into wood.

He saw a sparrow hawk speeding along the edge of the pond to a nameless destination. A male ruffed grouse burst out of deep snow with a stutter of powerful wings and drew a straight line of flight through the forest south of the pond. This mottled sweep of birches, beeches, and maples was silent and desolate. But the red-tailed hawk had seen flowers in its clearings and heard a score of bubbling, piping, whistling, chuckling cries of creatures living there in another season. He looked now and saw irradiant snow in a shaft of sun.

He wheeled over the marsh, which had only vestigial traces of the welling life it had held. Mink trails left shadows in the snow, fox prints crossed rabbit tracks, mice paths ran into snowbanks, and watching muskrats were black marks beside their shelters.

A black and white downy woodpecker flew over the pond and began drumming for borers in the hollow stems of horseweed stalks gripped in ice. The brittle sound of his smashing beak sent a flock of redpolls scattering across the ice. An undulating charm of greenish-yellow goldfinches flew noisily over the pond and swerved down toward the angular wreckage of thistles standing up out of the snow.

The hawk turned slowly and flexed his great wings to maintain his height. In this cold wind, he flew merely to see and to travel. Gone was the exhilaration of fast-rising summer air carrying him so high into the sky's blue vacuum that the pond became a silver speck and the great southern lake dazzled him with a glaring slash of reflected sun. Gone also were the fall gales that could tumble him in massive turbulences of cold air meeting warm.

There was no urge now to scream with an excitement that would give pause to every ground creature hearing him. His hunting was so lean that all his senses were gripped by the need for food. He was trapped in a hostile winter, enduring it with strength acquired in another season, when blackbirds flashed red shoulders in the marsh, when ducks dropped low over the trees and smashed the sheen of the pond into dancing reflections, when rabbits and hares came down through the northern forest of oaks, hickories, and elms, passing from clearing to clearing and eventually disappearing into the dense overgrowth of water plantains, alders, sedges, and rushes that bordered the pond. The red-tailed hawk looked at the northern forest now but saw only a late sun coloring the snow.

The snowshoe hare knew another dimension of the pond. He waited by the ice, his breath fogging the air. He waited, watching the red-tailed hawk pivoting slowly overhead. He waited, smelling a fox. The hawk turned to the south, and the fox appeared from a thicket, plowing through chest-high snow and then disappearing into the shadows of the forest. The hare rose slowly from his crouch. Trembling, he sniffed.

The pond to him was an arena of dangers, an enclosure of leaning dark trees whence came ticks and whistles and scuffs of alarming sound. He trusted nothing. He had seen a pair of goshawks—great northern birds as large as he—floundering through the snow and brush on the tracks of cottontail rabbits and had heard screams when the rabbits were caught. Once, he had been watching a dog fox running across the far side of the pond when the bitch had leaped at him from the brush. He had hurled himself onto the pond, his bristled feet scrabbling for a grip on the slick ice as the bitch chopped her teeth close behind him.

Being lighter than the fox, he had gained speed faster and shot up the northern slopes ahead of explosions of kicked snow.

The hare watched the woodpeckers drumming for borers and saw a shower of pine-cone petals falling from a tree top where a flock of crossbills were using their strangely shaped beaks to wrench seeds from cones. The hare could see, but not comprehend, other birds finding and eating spider sacs and moth cocoons that hung on twigs and bagworm cases that swung in the wind. He watched the woodpeckers driving their beaks into tree trunks and pulling out sleeping grubs. He saw the mice sunning themselves at the entrances of their snow tunnels, which riddled parts of the snow around the pond. He saw an ermined weasel whisk into a snow bank and heard the terrified squeaks of mice desperately scuttling through collapsing snow corridors as the weasel thrust after them.

He saw the dark, rigid shape of a dead crow standing on a branch in a nearby tree. On first seeing the crow in early winter, the hare had been wary, but later he had realized that the black shape was harmless. He could not know that the young and inexperienced bird had been caught in an exposed roosting place one night by a bitingly cold wind. The crow had wakened and looked over the moonlit pond through bleared eyes and felt a paralyzing lassitude running through his body. He wanted to fly, but was held back by his fear of the dark. His powerful claws were locked on the branch in their sleeping grip, and he had died in the midmorning hours of darkness while slowly settling back on his perch. The morning sun flashed in his open eyes, and the hare bounded lightly away.

Like the hawk, the hare lived in a pungent aftermath of a beneficent season. South of the pond, he could see a great shoulder of granite with satellite rocks scattered around it, and in warm seasons he would lope among the rocks and cautiously crop succulent grasses. West of the pond, he could see a long ridge of snow rising from the ice. This was a rocky, sandy finger of land that jutted from the bank near where the inlet stream spilled into the pond. To reach patches of herbage on the peninsula, the hare might leap over the stream and graze, surrounded by the buzz and burr of thousands of crickets who lived in the sandy soil.

The hare, being a year-round resident, knew many of the ebbs and flows of pond life. He had seen thousands of creatures rising out of the water and spreading by wing and foot through forest and marsh. He had crouched in fear when forty thousand wings beat over the pond in a mass. He had seen hordes of dying creatures falling from the sky. He had heard singing mice and seen flying spiders. His senses were acutely tuned to the tempo of life at the pond because he was one of the hunted.

Looking over the petrified pond now, he saw a blunt, grizzled head appear out of a small hole in the ice. The muskrat stood erect and sniffed the air. He saw the rabbit and ignored him. He also saw a weasel and slipped back into the water. The hare bounded silently through the northern trees, and the weasel was alone at the pond.

The muskrat swam under the ice toward his shelter near the sandy peninsula. A small fish fled into deep gloom. The muskrat was an old animal, graying at the muzzle, and his body was scarred by many escapes from hunters. A fight with a mink had left a long crescent-shaped scar on his right shoulder. Most of his right ear had gone, torn loose by an arctic owl three seasons before. He swam into shallowing water, past sticks and debris hanging from the ice, through odd effulgences of glowing light.

He swam over black slopes of mud and meadows of dark green grasses and past ribbons of old vegetation that soared to the ice roof and disappeared into it. He swam under a cluster of water beetles gathered around a bubble of air under the ice roof. He swam over semi-dormant dragonfly nymphs, the armor-clad aquatic youngsters of delicate flying creatures, and past a heap of leaves from which protruded two forked tails of fish that were drowsing the winter away.

For all his keen vision, the muskrat saw only an infinitesimal fraction of the microcosm of the pond. He saw worms so small they were whitish specks to his eyes, but he could not see the creatures the worms were pursuing and eating. His eyes caught only glimpses of the behemoths of this miniature world of the pond and missed the lilliputians, a million of which might not cover one of his eyes.

He swam on through this microscopic universe, unable to see the disaster that was decimating its creatures. The rotting vegetation and leaves on the bottom of the pond were exhausting the oxygen, and millions of creatures were suffocating. Even the fish were finding it hard to breathe, and some were dying.

The muskrat climbed a crawlway into the tumbled reeds of his shelter till he was above water level and inside a large dry chamber lined with leaves. He shook himself and lay down.

As he slept, perhaps he had a vision of easier times at the pond.

The reeds would be thick and green, and the lambent air would be filled with floating seeds and flying insects. He would stand erect on a stone, watching dragonflies sweeping overhead, seeing the red-tailed hawk hovering high above them and hearing the movements of other muskrats among the reeds, some of them digging for shellfish, others pulling cattail roots from the mud. He would be fat, his pelt glistening sleekly in the sun, and his flattened tail would be instantly ready to hit the water as a signal of danger. At dusk he would watch for owls and for predatory mink, who roamed through the marsh. He would swim deeply through the pond and see sunfish and bass and dace moving among aquatic plants. He would see lizardlike salamanders and aquatic worms and snails and huge beetles. He would see hundreds of nymphs, the immature forms of insects who would later leave their drab aquatic bodies and fly above the pond. He would eat some of these nymphs, and as he crunched the tough exterior of their bodies, a kingfisher might smash into the calm water overhead in a shower of stars. The muskrat would see thousands of creatures heading out of the water as the sun streamed greenly down or as the moon halted at the surface. He would see a few of the millions of eggs that would be dropped into the pond. He would, in a long summer, see much of the pond's life. He slept now, remote from the fact and fantasy of abundant seasons, and the frozen pond lay silently around his rough shelter.

In the middle of the winter equinox, the weather warmed briefly and the trees ran with water. Horseweeds faded spectrally into a rising mist, and a black squirrel kicked up a shower of snow as he raced to a tree. A pair of crows looked down somberly from an empty elm, and the red-tailed hawk dropped near a thicket and then flew by the marsh carrying a limp body. On the pond, a mink sniffed the muskrat's entrance in the ice, while the bitch fox watched her from the trees. A weasel glided over a hummock of snow and merged into the whiteness of the north bank.

As suddenly as it had started, the thaw was halted by a returning wave of arctic air. The warmer, lighter southern air was lifted under the wedge of the approaching cold, which rolled over the pond, dissipated the fog, and froze the trees. In the next flaming red dawn, the trees were shimmering and flashing like a phantasmagoria of prisms. The ice threw off multifarious hues—ambers, blues, pale reds—and these were broadcast so profusely by the red sun that for a day the pond seemed transformed from frozen immobility to glowing and sparkling motion.

But the winter marched on. That night, it destroyed the color. At dawn, the watery sun revealed motionless dark trees waiting for the spring.

The pond was the center of a universe that no one creature could comprehend. The red-tailed hawk knew some of it. He had hunted there in three seasons, but his knowledge was limited to one level of life at the pond. The old muskrat, who had survived six years there, knew another small part of it. A pair of arctic owls, who arrived at the pond as refugees from the really bitter weather of the far north, and brought with them an impression of white tundra, caribou herds, and the howling of wolves, had some special knowledge of it. The hunted hare knew fear in all the pond's seasons, but his life was only a fraction of this complex cosmos.

These creatures had neither the time nor the instinct to know all the incredible pond. The muskrat slept in his shelter. The rabbit crouched down in deep snow and the suffocation spread under the ice. The hawk turned the flat of his wings to the wind and sped south in a grand parabola and the pond dwindled into the distance.

THE SLEEPING BILLIONS

The creatures suffocating in the pond could not escape; they were mute victims, and they were also a small minority compared with those who had found, through countless millennia of experience, that the safest way to survive the winter was to sleep.

A strange sleep it was. For some, it was a drowsiness always close to waking. The fish, buried among leaves and hiding between stones, made spasmodic efforts to hunt for food, but the bulk of the sleepers slept so abysmally they seemed dead.

Most of them were invisible. They blanketed mud and sand slopes at the bottom of the pond and clustered under the snow, sheltered in crevices, under bark, stones, leaves, and logs. They were swathed in mud and buried in the earth itself. They were present everywhere in stupefying numbers. A pinch of soil might enfold a million of the simplest animals on earth, the single-celled protozoans, and scores of thousands of the simplest plants, or algae, and a million parasitic plants, or funguses, and millions of bacteria.

Some of these sleepers were frozen solidly into ice or ice-crystalled soil. All had reduced their physical demands to a slow pulse or near to zero. Millions of them slept inertly as eggs, and there were eggs everywhere: stuck to submarine plants, concealed in sand, mud, and wood, laid under rocks and stones. These were tough, overwintering eggs laid by creatures who could not survive winter as adults. Some animals endured as invisible blobs of jelly. Others, equally small, were encased in brittle, hard shells, or loricae. At this level of life, the division between plant and animal life was often indistinct or contradictory. Some invisible plants would awaken to live like animals, moving purposefully and rapidly through the water, and some animals would live passive, plantlike lives.

Some of this multitude would crawl on simple feet, swim with primitive arms, or laboriously build portable shelters out of granules of sand. Some would look like thunderclouds, or like ducks, or like tiny sparkling suns. Some would live in colonies big enough to color the water in many brilliant hues; others would live independently. Some were so small it might take them years to move from the bottom to the surface of the pond. The movements of some were controlled by the sun. Many were indescribable. One sleeping animal, which even in clustered thousands was scarcely discernible, was intricately constructed of innumerable minute hexagons, each one capable of radiating prismatic colors.

All these creatures—never visible to a raccoon's eye—were complete, self-contained individuals. They would, in the seasons to come, move their brainless, sightless protoplasm in endless fighting, hunting, and eating.

This diminutive host now slept, merged into the muck of the pond bottom.

The pond was a dormitory, with creatures sleeping above and below the snow and ice. The rocks, sand, earth, swamp, rotting heaps of summer vegetation, deadwood, and mud of the pond were refuges for an incoherent agglomeration of sleepers. The ground surrounding the pond was tunneled by earthworms. Any creature who dug to deposit eggs or to bury food or build a nest broke through several of these corridors. They were numerous enough to aerate the soil, drain it, make it more friable and a richer, surer stimulus for the growth of trees and plants. Most of the tunnels ran horizontally at shallow depths, but some, dug in the late fall, plunged down deeply. Two, three, or four of these would converge into sleeping galleries, or hibernacula, and were filled with convoluted masses of sleeping earthworms. The solitary earthworm, who used only the shallow corridors he dug, needed protection from both the frost of the surface and the dryness of the subsoil. So he co-operated with his fellows to dig deeper tunnels and spent the winter in moisture-retaining colonies. The smaller burrows contained fifty to sixty earthworms; the larger ones, more than one thousand. About five hundred million earthworms were asleep round the pond.

Despite the depth of their burial and their immobility, the earthworms could not entirely escape the dangers of the season. A mole dug down the line of one worm tunnel under a tree and eventually broke into a hibernaculum containing three hundred and fifty sleeping worms.

For most sleepers, the biggest danger was freezing. Thousands of carpenter ants had insulated themselves from this by burrowing deeply into a dead pondside maple. For years, the ants had drilled corridors through the trunk which connected and interconnected, branched and rebranched till they made the interior of the wood porous. The ants slept bunched together in long, black motionless lines in these corridors, and piled into heaps in the galleries that punctuated the tunnels.

They slept in a self-created microclimate, perfectly insulated from the freezing outside.

Other types of ants did not, or could not, escape the freezing. Instead, they had developed physical resistance to it, perhaps by anti-freezing their body liquids, perhaps by partially dehydrating themselves before the freeze began. A snow-covered, bulky ant mound near the north bank of the pond was deeply frosted through its corridor-riddled earth. In places, earth and ants were frozen into large crystalline masses. Below the frost line, the mound was flooded with pond water and some ants were asleep in it. In the narrow space between frost and water, dense masses of ants were piled together, jamming corridors and galleries in glutinous black clots. Their metabolism was so low that their hearts were motionless, and they lived in a secret suspension puzzlingly remote from the hot activity of their waking lives.

The ants were deep sleepers. They shared this quality with the ladybirds, small orange-colored beetles with black-spotted wing covers who slept in faults and crevices that riddled the great chunk of rock on the pond's south bank. One hundred and twenty thousand ladybirds piled together deep in one crevice. They returned to this hibernating place every year, and many rock crevices were filled with the decayed remains of scores of thousands of them who had not survived previous hibernations. Like the ants, they could endure freezing and submersion. Some rain water had seeped into the biggest mass of them, and several thousand were encased in clear ice. They were still alive, and most would survive to the thaw.

But few other creatures could long survive freezing. During periods of prolonged cold, the frost line deepened and enveloped creatures with low frost resistance. In the shallows, the water was frozen to the mud and the ice had reached down and gripped some sleeping frogs. They might survive if the ice did not reach their hearts. Like all hibernating frogs, they sustained themselves by drawing in minute quantities of oxygen through the skin.

The frogs in the mud of deeper water were safe. Chorus frogs were buried among the tiny green spring peepers; cricket frogs lay with leopard frogs; spotted frogs lay around a lone bullfrog who had buried himself much deeper than the others. He dwarfed them all in size and strength and was the last survivor of a dozen bullfrogs who had once lived at the pond but who had been wiped out by a pair of herons nesting there one year.

The frogs would soon overflow the pond were it not for some other sleepers who had a considerable appetite for frogs. A hundred snakes hunted at the pond, though in thick grass, reeds, and water their numbers were never noticeable. Green snakes, racers, and rat snakes, and handsome red, yellow, and black king snakes slept around the pond. There were speckled water snakes and garter snakes, red-bellied snakes, some deadly copperheads, and constricting bull snakes. They would pervade the life of the pond. But now they slept, their black eyes closed, and their bodies coiled coldly in heaps of leaves in hollow trees, under logs, and in ground-hog burrows, in old bumblebee nests, and among crevices and slits in the granite rock.

Two thousand turtle eggs were buried in mud, sand, and earth. Scores of buried painted-turtle eggs waited for a touch of warmth to hatch in strips of sand at the edge of the pond. The adult turtles, each as big as a sparrow, had buried themselves in mud nearby. Map turtles with yellow-lined shells, some twice as big as the painteds, slept among showy box turtles and spotted turtles. The pond and its littoral held one hundred and fifty turtles, some as long as a robin's outstretched wings, others smaller than a hummingbird.

The salamanders, who looked like lizards but had froglike skins and were nocturnal amphibians, equaled the turtles in variety and excelled them in colors. They were hidden in every likely winter refuge. The biggest of them were the deep brown waterdogs, who had bushing red gills and were each as long as a duck's wing. There were eastern newts, with yellow bellies and red spots on their green backs; tiny red efts who were smaller than chickadees; and spotted salamanders who had solid, chunky bodies and bright yellow spots on their dusky black backs. The salamanders were shaded green, brown, yellow, and red— which, with assorted spots, were their base colors. They spent the winter in every possible stage of development. Some lived as slimy clots of eggs. Others lived as gilled larva and looked like fish except for their rudimentary legs. Some, in adult form, seemed near death, so deep was their hibernation. A few remained active under the ice.

The survival of the cold-blooded animals was not as surprising as the endurance of more fragile creatures who, when awake, lived intensely and fleetingly. These lived and slept almost everywhere *except* in the pond. The remains of a hornets' nest bulked gloomily over the snow in the upper branches of a dogwood bush near the sandy peninsula. It was a three-tiered complex covered by a gray, waterproof material.

Inside were forty-eight dead hornets, still locked in the cells from which they had failed to escape in the late summer. Five thousand hornets had been born in this nest the previous year. The purplish-blue workers had whisked away across the pond, while the heavy-bodied drones hovered at the entrance of the ball-shaped nest. The workers had become drowsy and bedraggled in the fall and sought hiding places to die. All the hornets, except one, were now dead, and their bodies long since scattered into soil and water. The queen hornet, fertilized by a worker, had flown to the rock on the south bank to find a dry, deep crevice. Within her hunched, motionless body were six thousand separate lives of a future nest.

She shared the crevice with another sleeping queen, a large and rotund bumblebee, herself the only survivor of a nest of insects. She had been fertilized the previous fall also and would be among the first to leave this sleeping place and establish a new colony in an underground hive. By late summer, she might build it up to seven hundred bees.

A solitary overwintering queen was an efficient means for a colony to survive the winter, but it was not the only way. Several families of carpenter bees were asleep in the dead branch of an elm overhanging the pond. They had returned to sleep in tunnels cut into the wood by their parents or by themselves during the previous summer. They had developed there as larvae. They lay one behind the other, the head of one bee almost touching the abdomen of the bee in front. Unlike the solitary queens across the pond, the females were not fertilized. One bee in ten was a male. This percentage was just enough to ensure fertilization of the females on waking.

Nearby, in dead goldenrod stems in the swamp, smaller, metallic-blue carpenter bees slept in similar tunnels cut through the soft pithy stalk centers. They slept head downward, in file and abreast, in places packed in so closely that the earliest wakers might be impeded in their way to the new season.

Also in the swamp were sleeping wasps, bumblebees, spiders, and beetles, hidden in old nests in the wreckage of the cattails. None felt the vibration of blows from questing beaks or heard the harsh clatter of dry stalks on windy days. They were insensible to the sibilant scratching of feet seeking a grip on stalks and to the soft throat noises of hungry birds communicating in the biting cold air outside.

No common equation of survival united the sleepers. They slept naked and clothed, buried and exposed, as adults and as eggs. Many flies, like bluebottles and greenbottles, unable to survive freezing, had pushed deeply into the porous insulation of rotten branches and logs. Others clustered in hollow trees. The flies were immensely varied and incredibly numerous; their larvae permeated water and leaves, and their eggs lay in bark and under stones. When the flies began emerging from their sleep, the pond would sound to the roar, rasp, whine, screech, drone, and rumble of their wings.

Most butterflies and moths slept in a cocoon or chrysalis stage of life in which they were changing, inside dark-colored tubelike cases, from caterpillars to butterflies, in a mysterious pupation process. Luna moths lay in their cocoons in leaves and summer wreckage under the snow, and a field mouse, digging a snow tunnel, found one in some leaves and ate it. The luna caterpillars had wrapped themselves in leaves on the forest floor in the fall and were pupating there. Huge polyphemus moths, tightly cocooned in twisted leaves, lay scattered about the forest ground with the lunas.

Some pupae of butterflies slept in chrysalis cases that jutted from twigs and closely resembled the twigs in appearance. Great promethea moths slept as cocooned grubs in sagging, wrapped leaves in a grove of cherry trees. Red-admiral cocoons hung in ball-shaped houses, but for some reason as obscure as the hidden process that was so radically transforming this cocooned life, many red-admiral butterflies slept as adults in rotten logs and even under piles of leaves. Mourning cloak butterflies also slept as adults, their yellow-fringed wings spread in gloomy crevices of the rocks.

This intermediate stage of life survived the winter despite the fact that many chrysalises hung in full view of all the pond, suspended from twigs, branches, saplings, and hanging in piles of dried vegetation. They seemed to be the most vulnerable sleepers, and perhaps they were, for their mortality rate was high. Most mice, voles, woodpeckers, and chickadees hunted them eagerly.

A downy woodpecker came flatly across the pond into an old oak late one afternoon, saw a moth cocoon hanging at the far end of a fine-stemmed branch. He clutched clumsily down the slender stem and managed to peck the cocoon twice. But it swung around sharply on its retaining cord so that the blows had no effect on its tough cover. Eventually, the woodpecker flew away and the pupa inside the cocoon slept on.

The diverse, omnipresent beetles were other sleepers of boundless ingenuity, aggressiveness, and activity, who, once they were awake, would dig, bore, fly, swim, be cannibalistic, herbivorous, and carnivorous. They had in common armored skins, six legs, and a range of brilliant colors. Painted hickory borer beetles slept in long tunnels in a grove of hickories. Innumerable locust borers, in larval form, slept in shallow galleries just under the bark of trees. The big sugar-maple borer beetles, handsomely black and yellow, slept behind the frozen sides of nearly every one of the five hundred sugar maples round the pond.

There were beetles hidden in woodpecker borings and buried in the sunken buds of pine trees, in hollow stalks and in rock cracks, and in disused carpenter-ant borings. Many of them had gone into the soil to escape the cold, and underground the May beetle larvae were fifty times their own length. Even at that depth, there was a hint of the quality of life that was soon to spread over the pond. Among the sleeping beetles were hibernating pupae of black digger wasps. These had reached the earth the previous summer as eggs laid by female digger wasps who had dug into the soil to find the beetle grubs. Their eggs had hatched into larvae, which had eaten many of the grubs, had pupated, and were now waiting to develop into winged creatures for the summer.

The obvious sleepers were the animals and insects. Less obvious were the sleeping seeds. There was no way of estimating their numbers. At all levels of the pond's life—in the soil, water, and mud—seeds were waiting. Some had waited years to germinate and so were the longest, deepest sleepers of all. The hidden ground was absorbing a great many of them, including acorns, nuts, cones, single- and double-winged seeds of many types, seeds from berries and fruits. Many of them had special methods of warding off the teeming funguses and bacteria all around them and of awaiting a coincidence of events that would give them visible life.

A grove of elm trees the previous fall had liberated ten million winged seeds, which had rotated across the pond in chattering columns. They had fallen in water and on rock and had piled in drifts against banks. They represented the most profligate waste of pond life, because not more than a dozen of them would ever germinate. And the chances that these would grow into trees were remote. The pond was an ordered community of life, and the start of new life demanded first the destruction of the old. The seeds remained quiescent, but the trees would keep flying new crops of them over the pond in a ceaseless testing of the remotest chances to start new life.

All the trees were sleepers, even the evergreen white pines, spruces, larches, firs, hemlocks, and cedars, which kept their needled foliage throughout the year. Few other living things around the pond existed with such energy or with such insatiable demands on the resources around them. In their waking lives, they sucked incalculable amounts of water from the ground and extended roots and hair roots through the soil with such speed and power that subterranean creatures could sense the movement all around them. Cicada grubs would find hair roots entering their burrows, appearing out of the damp, glazed walls with almost perceptible speed. All this activity was now stopped, and the roots waited for the stimulus of spring to send them surging into a cohesive chain of movement to animate the highest branches above them. But now, the life force in them was collapsed to its lowest level.

A colony of bats slept in the deepest crevice in the granite rocks, and their sleep was unique. In gray-tinted darkness, they hung in hundreds from the sloping roof, so closely packed together that the rock was concealed by their black, podlike forms. One bat was not breathing, and long moments passed before he took one shallow intake of air. He breathed ten times in an hour. Another bat was breathing once every two seconds. The breathing of a third bat began increasing rapidly. First she took one breath every second; then two breaths a second; then three, till her furred sides were pulsing in a blur. Her eyes opened and she stretched. Her activity seemed communicated to the surrounding bats, and their breathing quickened. The first bat silently dropped and flitted through the gloom, and others soon followed, till there were fifty or sixty of them darting up and down the narrow crevice, their wings whispering softly but never touching as they passed and repassed. Then, as quickly as the flying had begun, it ended. The bats swooped up to the top of the cave, fastened themselves, licked their fur and wings briefly, and instantly relapsed into the drowsing of hibernation. The gloom of their hibernaculum contrasted oddly with the harsh light of the sinking sun outside, where the snowshoe hare crouched in the silent snow. Near him, a mass of ice crystals at the entrance of a hollow tree revealed the presence of other sleepers. The crystals grew from a continuous escape of warm air from two sleeping raccoons. The male lay on his back, curled up, with his eyes tightly covered by both paws. The female had jammed herself into a narrow space beside him and slept on her head, with her furry posterior pointing upwards.

Like the bats, the raccoons tended to be intermittent sleepers. They would soon rouse themselves and go into the snow to mate, and then resume their winter sleep. Their exceptional fatness, built up before they went into the winter sleep, sustained them, and their slowed-down bodies drew on this fat sparingly. A family of skunks soundly asleep in a burrow near the bank of the pond used the same sustaining device. They were seen occasionally venturing with distaste through the deep snow.

Some sleepers lived on stored food. The chipmunks, asleep in deep underground burrows, had dug storage chambers that they had packed full of acorns, nuts, seeds, and grass hay in the fall. They would rouse themselves occasionally, take a brief meal in the storage chambers, and visit a relief chamber where they emptied bladders and bowels before returning to the sleeping chamber to resume hibernation.

All the sleeping creatures used mud, earth, wood, leaves, snow, ice, or water as insulation against the cold. But some creatures built their own insulation or carried it with them. Spiders endured as eggs or as dormant youngsters in egg cases, or as adults. Several thousand spiderlings and eggs waited under a rock face, perfectly hidden to outlast any weather. Some adult spiders had infiltrated piles of leaves and spun themselves into winter shelters of the finest and warmest silk.

The exceptional cold tested the insulation of all these materials during this winter. The deep snow and the bitter temperatures were gradually killing many creatures unprepared for such severity. The wind swept snow from one exposed shoulder of land on a northern plateau, and the earth, deprived of this cover, froze deeply. The ice was spreading through soil and leaf mould, and although some twenty thousand hibernating ladybirds there would survive this onslaught, beetles, flies, butterflies, and bees, sleeping contiguously, were dying.

It was a partial sleep for some. The old muskrat dozed in his pond shelter during the days. But down in the marsh, many muskrats had been made stuporous by the cold; the hunting mink swam under the ice, entered their shelters, dragged them into the snow, and killed them. The muskrats struggled feebly or not at all, their resistance to death broken by the cold.

For others, like the red-tailed hawk in the branches of a bare oak, it was an uneasy and intermittent sleep. The hawk awoke and shivered violently, and these spasms of his muscles helped warm him a little. For the arctic owls, it was a warm sleep by day and night behind dense-feathered plumage.

The cold had caught several thousand aquatic snails in its solid clutch in the northern shallows of the pond. The snails could be seen clearly through the ice, fully withdrawn into their shells and sustaining themselves on tiny amounts of oxygen drawn through the ice. But this was a diminishing supply, and hundreds of them had already died.

The obscure dying of the snails was akin to another death by suffocation in the deeper waters of the pond. About one hundred fish—carp, sunfish, and dace—had died as oxygen in the pond diminished; their bodies rose out of leaves and mud and slowly ascended to the ice.

The numbers of the sleeping creatures were now evident. How could the pond cope with their awakening? What could prevent chaos as the sleeping billions stirred into life and began filling every part of the pond—its soil, mud, air, and water? What cosmic force would wring order and continuity out of a massive host of divergent life forces thrusting and straining in every direction? How could there always be a balance of these lives when their strengths and numbers varied so sharply? Why did not one creature—witness the layers of sleeping ants, or the clots of gallery-bound earthworms, the bulk of ladybirds—come to dominate the pond's life in a triumph of a single species? What was the purpose of such a multitude of species?

The sleepers, lying under the quiet snow, in gloomy water and constricting mud, posed many questions, and the answers to them all still did not reveal the secret of the pond.

THE RETURN
OF THE BIRDS *

JOHN BURROUGHS

John Burroughs, like many of the early naturalists, was a scientist, poet, philosopher, and a rugged outdoorsman. He was greatly influenced by nature-poet Ralph Waldo Emerson and the great individualist, Henry David Thoreau.

His books include "The Ways of Nature," "Accepting the Universe," "Squirrels and Other Fur-bearers," "Locusts and Wild Honey," and "Birds and Poets."

* From "Wake-robin" by John Burroughs. Copyright © 1871, 1876 by John Burroughs. Copyright © 1885 by Houghton Mifflin & Co. Reprinted by permission of Houghton Mifflin Company.

S pring in our northern climate may fairly be said to extend from the middle of March to the middle of June. At least, the vernal tide continues to rise until the latter date, and it is not till after the summer solstice that the shoots and twigs begin to harden and turn to wood, or the grass to lose any of its freshness and succulency.

It is this period that marks the return of the birds,—one or two of the more hardy or half-domesticated species, like the song-sparrow and the bluebird, usually arriving in March, while the rarer and more brilliant wood-birds bring up the procession in June. But each stage of the advancing season gives prominence to certain species, as to certain flowers. The dandelion tells me when to look for the swallow, the dog-toothed violet when to expect the wood-thrush, and when I have found the wake-robin in bloom I know the season is fairly inaugurated. With me this flower is associated, not merely with the awakening of Robin, for he has been awake some weeks, but with the universal awakening and rehabilitation of nature.

Yet the coming and going of the birds is more or less a mystery and a surprise. We go out in the morning, and no thrush or vireo is to be heard; we go out again, and every tree and grove is musical; yet again, and all is silent. Who saw them come? Who saw them depart?

This pert little winter-wren, for instance, darting in and out the fence, diving under the rubbish here and coming up yards away, how does he manage with those little circular wings to compass degrees and zones, and arrive always in the nick of time? Last August I saw him in the remotest wilds of the Adirondacs, impatient and inquisitive as usual; a few weeks later, on the Potomac, I was greeted by the same hardy little busybody. Does he travel by easy stages from bush to bush and from wood to wood? or has that compact little body force and courage to brave the night and the upper air, and so achieve leagues at one pull?

And yonder bluebird with the earth tinge on his breast and the sky tinge on his back,—did he come down out of heaven on that bright March morning when he told us so softly and plaintively that if we pleased, spring had come? Indeed, there is nothing in the return of the birds more curious and suggestive than in the first appearance, or rumors of the appearance, of this little bluecoat. The bird at first seems a mere wandering voice in the air; one hears its call or carol on some bright March morning, but is uncertain of its source or direction; it falls like a drop of rain when no cloud is visible; one looks and listens, but to no purpose. The weather changes, perhaps a cold snap with snow comes on, and it may be a week before I hear the note again, and this time or the next perchance see the bird sitting on a stake in the fence lifting his wing as he calls cheerily to his mate. Its notes now become daily more frequent, the birds multiply, and, flitting from point to point, call and warble more confidently and gleefully. Their boldness increases till one sees them hovering with a saucy, inquiring air about barns and out-buildings, peeping into dove-cotes, and stable windows, inspecting knot-holes and pump-trees, intent only on a place to nest. They wage war against robins and wrens, pick quarrels with swallows, and seem to deliberate for days over the policy of taking forcible possession of one of the mud-houses of the latter. But as the season advances they drift more into the background. Schemes of conquest which they at first seemed bent upon are abandoned, and they settle down very quietly in their old quarters in remote stumpy fields.

Not long after the bluebird comes the robin, sometimes in March, but in most of the Northern States April is the month of the robin. In large numbers they scour the fields and groves. You hear their piping in the meadow, in the pasture, on the hill-side. Walk in the woods, and the dry leaves rustle with the whir of their wings, the air is vocal with their cheery call. In excess of joy and vivacity, they run, leap, scream, chase each other through the air, diving and sweeping among the trees with perilous rapidity.

In that free, fascinating, half-work and half-play pursuit, sugar-making,—a pursuit which yet lingers in many parts of New York, as in New England,—the robin is one's constant companion. When the day is sunny and the ground bare, you meet him at all points and hear him at all hours. At sunset, on the tops of the tall maples, with look heavenward, and in a spirit of utter abandonment, he carols his simple strain. And sitting thus amid the stark, silent trees, above the wet, cold earth, with the chill of winter still in the air, there is no fitter or sweeter songster in the whole round year. It is in keeping with the scene and the occasion. How round and genuine the notes are, and how eagerly our ears drink them in! The first utterance, and the spell of winter is thoroughly broken, and the remembrance of it afar off.

Robin is one of the most native and democratic of our birds; he is one of the family, and seems much nearer to us than those rare, exotic visitants, as the orchard starling or rose-breasted grossbeak, with their distant, high-bred ways. Hardy, noisy, frolicsome, neighborly and domestic in his habits, strong of wing and bold in spirit, he is the pioneer of the thrush family, and well worthy of the finer artists whose coming he heralds and in a measure prepares us for.

I could wish Robin less native and plebeian in one respect,—the building of his nest. Its coarse material and rough masonry are creditable neither to his skill as a workman nor to his taste as an artist. I am the more forcibly reminded of his deficiency in this respect from observing yonder hummingbird's nest, which is a marvel of fitness and adaptation, a proper setting for this winged gem, the body of it composed of a white, felt-like substance, probably the down of some plant or the wool of some worm, and toned down in keeping with the branch on which it sits by minute tree-lichens, woven together by threads as fine and frail as gossamer. From Robin's good looks and musical turn we might reasonably predict a domicile of equal fitness and elegance. At least I demand of him as clean and handsome a nest as the kingbird's, whose harsh jingle, compared with Robin's evening melody, is as the clatter of pots and kettles beside the tone of a flute. I love his note and ways better even than those of the orchard starling or the Baltimore oriole; yet his nest, compared with theirs, is a half-subterranean hut contrasted with a Roman villa. There is something courtly and poetical in a pensile nest. Next to a castle in the air is a dwelling suspended to the slender branch of a tall tree, swayed and rocked forever by the wind. Why need wings be afraid of falling? Why build only where boys can climb? After all, we must set it down to the account of Robin's democratic turn; he is no aristocrat, but one of the people; and therefore we should expect stability in his workmanship rather than elegance.

Another April bird, which makes her appearance sometimes earlier and sometimes later than Robin, and whose memory I fondly cherish, is the Phoebe-bird, the pioneer of the fly-catchers. In the inland farming districts I used to notice her, on some bright morning about Easter-day, proclaiming her arrival with much variety of motion and attitude, from the peak of the barn or hay-shed. As yet, you may have heard only the plaintive, homesick note of the bluebird, or the faint trill of the song-sparrow, and Phoebe's clear, vivacious assurance of her veritable bodily presence among us again is welcomed by all ears. At agreeable intervals in her lay she describes a circle or an ellipse in the air, ostensibly prospecting for insects, but really, I suspect, as an artistic flourish, thrown in to make up in some way for the deficiency of her musical performance. If plainness of dress indicates powers of song, as it usually does, then Phoebe ought to be unrivalled in musical ability, for surely that ashen-gray suit is the superlative of plainness; and that form, likewise, would hardly pass for a "perfect figure" of a bird. The seasonableness of her coming, however, and her civil, neighborly ways, shall make up for all deficiencies in song and plumage. After a few weeks Phoebe is seldom seen, except as she darts from her moss-covered nest beneath some bridge or shelving cliff.

Another April comer, who arrives shortly after Robin-redbreast, with whom he associates both at this season and in the autumn, is the gold-winged woodpecker, *alias* "high-hole," *alias* "flicker," *alias* "ya-rup." He is an old favorite of my boyhood, and his note to me means very much. He announces his arrival by a long, loud call, repeated from the dry branch of some tree, or a stake in the fence—a thoroughly melodious April sound. I think how Solomon finished that beautiful description of spring, "And the voice of the turtle is heard in the land," and see that a description of spring in this farming country, to be equally characteristic, should culminate in like manner,—"And the call of the high-hole comes up from the wood."

It is a loud, strong, sonorous call, and does not seem to imply an answer, but rather to subserve some purpose of love or music. It is "Yarup's" proclamation of peace and good-will to all. On looking at the matter closely, I perceive that most birds, not denominated songsters, have, in the spring, some note or sound or call that hints of a song, and answers imperfectly the end of beauty and art. As a "livelier iris changes on the burnished dove," and the fancy of the young man turns lightly to thoughts of his pretty cousin, so the same renewing spirit touches the "silent singers," and they are no longer dumb; faintly they lisp the first syllables of the marvellous tale. Witness the clear, sweet whistle of the gray-crested titmouse, the soft, nasal piping of the nuthatch, the amorous, vivacious warble of the bluebird, the long, rich note of the meadow-lark, the whistle of the quail, the drumming of the partridge, the animation and loquacity of the swallows, and the like. Even the hen has a homely, contented carol; and I credit the owls with a desire to fill the night with music. All birds are incipient or would-be songsters in the spring. I find corroborative evidence of this even in the crowing of the cock. The flowering of the maple is not so obvious as that of the magnolia; nevertheless, there is actual inflorescence.

Few writers award any song to that familiar little sparrow, the *Socialis;* yet who that has observed him sitting by the wayside, and repeating, with devout attitude, that fine sliding chant, does not recognize the neglect? Who has heard the snowbird sing? Yet he has a lisping warble very savory to the ear. I have heard him indulge in it even in February.

Even the cow-bunting feels the musical tendency, and aspires to its expression, with the rest. Perched upon the topmost branch, beside his mate or mates,—for he is quite a polygamist and usually has two or three demure little ladies in faded black beside him,—generally in the early part of the day, he seems literally to vomit up his notes. Apparently with much labor and effort, they gurgle and blubber up out of him, falling on the ear with a peculiar subtile ring, as of turning water from a glass bottle, and not without a certain pleasing cadence.

Neither is the common woodpecker entirely insensible to the wooing of the spring, and, like the partridge, testifies his appreciation of melody after quite a primitive fashion. Passing through the woods, on some clear, still morning in March, while the metallic ring and tension of winter are still in the earth and air, the silence is suddenly broken by long, resonant hammering upon a dry limb or stub. It is Downy beating a reveille to spring. In the utter stillness and amid the rigid forms we listen with pleasure; and as it comes to my ear oftener at this season than at any other, I freely exonerate the author of it from the imputation of any gastronomic motives, and credit him with a genuine muscial performance.

It is to be expected, therefore, that "Yellow-hammer" will respond to the general tendency, and contribute his part to the spring chorus. His April call is his finest touch, his most musical expression.

I recall an ancient maple standing sentry to a large sugar-bush, that, year after year, afforded protection to a brood of yellow-hammers in its decayed heart. A week or two before the nesting seemed actually to have begun, three or four of these birds might be seen, on almost any bright morning, gambolling and courting amid its decayed branches. Sometimes you would hear only a gentle, persuasive cooing, or a quiet, confidential chattering; then that long, loud call, taken up by first one, then another, as they sat about upon the naked limbs; anon, a sort of wild, rollicking laughter, intermingled with various cries, yelps, and squeals, as if some incident had excited their mirth and ridicule. Whether this social hilarity and boisterousness is in celebration of the pairing or mating ceremony, or whether it is only a sort of annual "house-warming" common among high-holes on resuming their summer quarters, is a question upon which I reserve my judgment.

Unlike most of his kinsmen, the golden-wing prefers the fields and the borders of the forest to the deeper seclusion of the woods, and hence, contrary to the habit of his tribe, obtains most of his subsistence from the ground, probing it for ants and crickets. He is not quite satisfied with being a woodpecker. He courts the society of the robin and the finches, abandons the trees for the meadow, and feeds eagerly upon berries and grain. What may be the final upshot of this course of living is a question worthy the attention of Darwin. Will his taking to the ground and his pedestrian feats result in lengthening his legs, his feeding upon berries and grains subdue his tints and soften his voice, and his associating with Robin put a song into his heart?

Indeed, what would be more interesting than the history of our birds for the last two or three centuries? There can be no doubt that the presence of man has exerted a very marked and friendly influence upon them, since they so multiply in his society. The birds of California, it is said, were mostly silent till after its settlement, and I doubt if the Indians heard the wood-thrush as we hear him. Where did the bobolink disport himself before there were meadows in the North and rice fields in the South? Was he the same blithe, merry-hearted beau then as now? And the sparrow, the lark, and the goldfinch, birds that seem so indigenous to the open fields and so averse to the woods,—we cannot conceive of their existence in a vast wilderness and without man.

But to return. The song-sparrow, that universal favorite and firstling of the spring, comes before April, and its simple strain gladdens all hearts.

May is the month of the swallows and the orioles. There are many other distinguished arrivals, indeed nine tenths of the birds are here by the last week in May, yet the swallows and orioles are the most conspicuous. The bright plumage of the latter seems really like an arrival from the tropics. I see them dash through the blossoming trees, and all the forenoon hear their incessant warbling and wooing. The swallows dive and chatter about the barn, or squeak and build beneath the eaves, the partridge drums in the fresh sprouting woods; the long, tender note of the meadow-lark comes up from the meadow; and at sunset, from every marsh and pond come the ten thousand voices of the hylas. May is the transition month, and exists to connect April and June, the root with the flower.

With June the cup is full, our hearts are satisfied, there is no more to be desired. The perfection of the season, among other things, has brought the perfection of the song and plumage of the birds. The master artists are all here; and the expectations excited by the robin and the song-sparrow are fully justified. The thrushes have all come; and I sit down upon the first rock, with hands full of the pink azalea, to listen. With me the cuckoo does not arrive till June; and often the goldfinch, the king-bird, the scarlet tanager delay their coming till then. In the meadows the bobolink is in all his glory; in the high pastures the field-sparrow sings his breezy vesper-hymn; and the woods are unfolding to the music of the thrushes.

The cuckoo is one of the most solitary birds of our forests, and is strangely tame and quiet, appearing equally untouched by joy or grief, fear or anger. Something remote seems ever weighing upon his mind. His note or call is as of one lost or wandering, and to the farmer is prophetic of rain. Amid the general joy and the sweet assurance of things, I love to listen to the strange clairvoyant call. Heard a quarter of a mile away, from out the depths of the forest, there is something peculiarly weird and monkish about it. Wordsworth's lines upon the European species apply equally well to ours:

> O blithe new-comer! I have heard,
> I hear thee and rejoice:
> O cuckoo! shall I call thee bird?
> Or but a wandering voice?

> While I am lying on the grass,
> Thy loud note smites my ear!
> From hill to hill it seems to pass,
> At once far off and near!

> Thrice welcome, darling of the spring!
> Even yet thou art to me
> No bird, but an invisible thing,
> A voice, a mystery.

The black-billed is the only species found in my locality, the yellow-billed abounds farther south. Their note or call is nearly the same. The former sometimes suggests the voice of a turkey. The call of the latter may be suggested thus: *k-k-k-k-k-kow, kow, kow-ow, kow-ow.*

The yellow-billed will take up his stand in a tree and explore its branches till he has caught every worm. He sits on a twig, and with a peculiar swaying movement of his head examines the surrounding foliage. When he discovers his prey, he leaps upon it in a fluttering manner.

In June the black-billed makes a tour through the orchard and garden, regaling himself upon the canker-worms. At this time he is one of the tamest of birds, and will allow you to approach within a few yards of him. I have even come within a few feet of one without seeming to excite his fear or suspicion. He is quite unsophisticated, or else royally indifferent.

The plumage of the cuckoo is a rich glossy brown, and is unrivalled in beauty by any other neutral tint with which I am acquainted. It is also remarkable for its firmness and fineness.

Notwithstanding the disparity in size and color, the black-billed species has certain peculiarities that remind one of the passenger-pigeon. His eye, with its red circle, the shape of his head, and his motions on alighting and taking flight, quickly suggest the resemblance; though in grace and speed, when on the wing, he is far inferior. His tail seems disproportionately long, like that of the red thrush, and his flight among the trees is very still, contrasting strongly with the honest clatter of the robin or pigeon.

Have you heard the song of the field-sparrow? If you have lived in a pastoral country with broad upland pastures, you could hardly have missed him. Wilson, I believe, calls him the grass-finch, and was evidently unacquainted with his powers of song. The two white lateral quills in his tail, and his habit of running and skulking a few yards in advance of you as you walk through the fields, are sufficient to identify him. Not in meadows or orchards, but in high, breezy pasture-grounds, will you look for him. His song is most noticeable after sundown, when other birds are silent; for which reason he has been aptly called the vesper-sparrow. The farmer following his team from the field at dusk catches his sweetest strain. His song is not so brisk and varied as that of the song-sparrow, being softer and wilder, sweeter and more plaintive. Add the best parts of the lay of the latter to the sweet vibrating chant of the wood-sparrow, and you have the evening hymn of the vesper-bird,—the poet of the plain, unadorned pastures. Go to those broad, smooth, uplying fields where the cattle and sheep are grazing, and sit down in the twilight on one of those warm, clean stones, and listen to this song. On every side, near and remote, from out the short grass which the herds are cropping, the strain rises. Two or three long, silver notes of peace and rest, ending in some subdued trills and quavers, constitute each separate song. Often you will catch only one or two of the bars, the breeze having blown the minor part away. Such unambitious, quiet, unconscious melody! It is one of the most characteristic sounds in Nature. The grass, the stones, the stubble, the furrow, the quiet herds, and the warm twilight among the hills, are all subtilely expressed in this song; this is what they are at last capable of.

The female builds a plain nest in the open field, without so much as a bush or thistle or tuft of grass to protect it or mark its site; you may step upon it or the cattle may tread it into the ground. But the danger from this source, I presume, the bird considers less than that from another. Skunks and foxes have a very impertinent curiosity, as Finchie well knows,—and a bank or hedge, or a rank growth of grass or thistles, that might promise protection and cover to mouse or bird, these cunning rogues would be apt to explore most thoroughly. The partridge is undoubtedly acquainted with the same process of reasoning; for, like the vesper-bird, she, too, nests in open, unprotected places, avoiding all show of concealment,—coming from the tangled and almost impenetrable parts of the forest, to the clean, open woods, where she can command all the approaches and fly with equal ease in any direction.

Another favorite sparrow, but little noticed, is the wood or bush sparrow, usually called by the ornithologists *Spizella pusilla*. Its size and form is that of the *socialis*, but is less distinctly marked, being of a duller, redder tinge. He prefers remote bushy heathery fields, where his song is one of the sweetest to be heard. It is sometimes very noticeable, especially early in spring. I remember sitting one bright day in the still leafless April woods, when one of these birds struck up a few rods from me, repeating its lay at short intervals for nearly an hour. It was a perfect piece of wood-music, and was of course all the more noticeable for being projected upon such a broad unoccupied page of silence. Its song is like the words, *fe-o, fe-o, fe-o, few, few, few, fee fee fee*, uttered at first high and leisurely, but running very rapidly toward the close, which is low and soft.

Still keeping among the unrecognized, the white-eyed vireo, or fly-catcher, deserves particular mention. The song of this bird is not particularly sweet and soft; on the contrary, it is a little hard and shrill, like that of the indigo-bird or oriole; but for brightness, volubility, execution, and power of imitation, he is unsurpassed by any of our northern birds. His ordinary note is forcible and emphatic, but, as stated, not especially musical: *Chick-a-re'r-chick*, he seems to say, hiding himself in the low, dense undergrowth, and eluding your most vigilant search, as if playing some part in a game. But in July or August, if you are on good terms with the sylvan deities, you may listen to a far more rare and artistic performance. Your first impression will be that that cluster of azalea, or that clump of swamp-huckleberry, conceals three or four different songsters, each vying with the others to lead the chorus. Such a medley of notes, snatched from half the songsters of the field and forest, and uttered with the utmost clearness and rapidity, I am sure you cannot hear short of the haunts of the genuine mocking-bird. If not fully and accurately repeated, there are at least suggested the notes of the robin, wren, cat-bird, high-hole, goldfinch, and song-sparrow. The *pip, pip* of the last is produced so accurately that I verily believe it would deceive the bird herself;—and the whole uttered in such rapid succession that it seems as if the movement that gives the concluding note of one strain must form the first note of the next. The effect is very rich, and, to my ear, entirely unique. The performer is very careful not to reveal himself in the mean time; yet there is a conscious air about the strain that impresses me with the idea that my presence is understood and my attention courted. A tone of pride and glee, and, occasionally, of bantering jocoseness, is discernible. I believe it is only rarely, and when he is sure of his audience, that he displays his parts in this manner. You are to look for him, not in tall trees or deep forests, but in low, dense shrubbery about wet places, where there are plenty of gnats and mosquitoes.

The winter-wren is another marvellous songster, in speaking of whom it is difficult to avoid superlatives. He is not so conscious of his powers and so ambitious of effect as the white-eyed fly-catcher, yet you will not be less astonished and delighted on hearing him. He possesses the fluency and copiousness for which the wrens are noted, and besides these qualities, and what is rarely found conjoined with them, a wild, sweet, rhythmical cadence that holds you entranced. I shall not soon forget that perfect June day, when, loitering in a low, ancient hemlock wood, in whose cathedral aisles the coolness and freshness seem perennial, the silence was suddenly broken by a strain so rapid and gushing, and touched with such a wild, sylvan plaintiveness, that I listened in amazement. And so shy and coy was the little minstrel, that I came twice to the woods before I was sure to whom I was listening. In summer he is one of those birds of the deep northern forests, that, like the speckled Canada warbler and the hermit-thrush, only the privileged ones hear.

The distribution of plants in a given locality is not more marked and defined than that of the birds. Show a botanist a landscape, and he will tell you where to look for the lady's-slipper, the columbine, or the harebell. On the same principles the ornithologist will direct you where to look for the greenlets, the wood-sparrow, or the chewink. In adjoining counties, in the same latitude, and equally inland, but possessing a different geological formation and different forest-timber, you will observe quite a different class of birds. In a land of the beech and sugar-maple I do not find the same songsters that I know where thrive the oak, chestnut, and laurel. In going from a district of the Old Red Sandstone to where I walk upon the old Plutonic Rock, not fifty miles distant, I miss in the woods the veery, the hermit-thrush, the chestnut-sided warbler, the blue-backed warbler, the green-backed warbler, the black and yellow warbler, and many others, and find in their stead the wood-thrush, the chewink, the redstart, the yellow-throat, the yellow-breasted fly-catcher, the white-eyed fly-catcher, the quail, and the turtle-dove.

In my neighborhood here in the Highlands the distribution is very marked. South of the village I invariably find one species of birds, north of it another. In only one locality, full of azalea and swamp-huckleberry, I am always sure of finding the hooded warbler. In a dense undergrowth of spicebush, witch-hazel, and alder, I meet the worm-eating warbler. In a remote clearing covered with heath and fern, with here and there a chestnut and an oak, I go to hear in July the wood-sparrow, and returning by a stumpy, shallow pond, I am sure to find the water-thrush.

Only one locality within my range seems to possess attractions for all comers. Here one may study almost the entire ornithology of the State. It is a rocky piece of ground, long ago cleared, but now fast relapsing into the wildness and freedom of nature, and marked by those half-cultivated, half-wild features which birds and boys love. It is bounded on two sides by the village and highway, crossed at various points by carriage-roads, and threaded in all directions by paths and by-ways, along which soldiers, laborers, and truant school-boys are passing at all hours of the day. It is so far escaping from the axe and the bush-hook as to have opened communication with the forest and mountain beyond by straggling lines of cedar, laurel, and blackberry. The ground is mainly occupied with cedar and chestnut, with an undergrowth, in many places, of heath and bramble. The chief feature, however, is a dense growth in the centre, consisting of dogwood, water-beech, swamp-ash, alder, spice-bush, hazel, etc., with a network of smilax and frost-grape. A little zigzag stream, the draining of a swamp beyond, which passes through this tangle-wood, accounts for many of its features and productions, if not for its entire existence. Birds that are not attracted by the heath or the cedar and chestnut are sure to find some excuse for visiting this miscellaneous growth in the centre. Most of the common birds literally throng this idle-wild; and I have met here many of the rarer species, such as the great-crested fly-catcher, the solitary warbler, the blue-winged swamp-warbler, the worm-eating warbler, the fox-sparrow, etc. The absence of all birds of prey, and the great number of flies and insects, both the result of proximity to the village, are considerations which no hawk-fearing, peace-loving minstrel passes over lightly; hence the popularity of the resort.

But the crowning glory of all these robins, fly-catchers, and warblers is the wood-thrush. More abundant than all the other birds, except the robin and cat-bird, he greets you from every rock and shrub. Shy and reserved when he first makes his appearance in May, before the end of June he is tame and familiar, and sings on the tree over your head, or on the rock a few paces in advance. A pair even built their nest and reared their brood within ten or twelve feet of the piazza of a large summer-house in the vicinity. But when the guests commenced to arrive and the piazza to be thronged with gay crowds, I noticed something like dread and foreboding in the manner of the mother-bird; and from her still, quiet ways, and habit of sitting long and silently within a few feet of the precious charge, it seemed as if the dear creature had resolved, if possible, to avoid all observation.

If we take the quality of melody as the test, the wood-thrush, hermit-thrush, and the veery-thrush stand at the head of our list of songsters.

The mocking-bird undoubtedly possesses the greatest range of mere talent, the most varied executive ability, and never fails to surprise and delight one anew at each hearing; but being mostly an imitator, he never approaches the serene beauty and sublimity of the hermit-thrush. The word that best expresses my feelings on hearing the mocking-bird is admiration, though the first emotion is one of surprise and incredulity. That so many and such various notes should proceed from one throat is a marvel, and we regard the performance with feelings akin to those we experience on witnessing the astounding feats of the athlete or gymnast, —and this, notwithstanding many of the notes imitated have all the freshness and sweetness of the originals. The emotions excited by the songs of these thrushes belong to a higher order, springing as they do from our deepest sense of the beauty and harmony of the world.

The wood-thrush is worthy of all, and more than all, the praises he has received; and considering the number of his appreciative listeners, it is not a little surprising that his relative and equal, the hermit-thrush, should have received so little notice. Both the great ornithologists, Wilson and Audubon, are lavish in their praises of the former, but have little or nothing to say of the song of the latter. Audubon says it is sometimes agreeable, but evidently has never heard it. Nuttall, I am glad to find, is more discriminating, and does the bird fuller justice.

It is quite a rare bird, of very shy and secluded habits, being found in the Middle and Eastern States, during the period of song, only in the deepest and most remote forests, usually in damp and swampy localities. On this account the people in the Adirondac region call it the "Swamp Angel." Its being so much of a recluse accounts for the comparative ignorance that prevails in regard to it.

The cast of its song is very much like that of the wood-thrush, and a good observer might easily confound the two. But hear them together and the difference is quite marked: the song of the hermit is in a higher key, and is more wild and ethereal. His instrument is a silver horn, which he winds in the most solitary places. The song of the wood-thrush is more golden and leisurely. Its tone comes near to that of some rare stringed instrument. One feels that perhaps the wood-thrush has more compass and power, if he would only let himself out, but on the whole he comes a little short of the pure, serene, hymn-like strain of the hermit.

Yet those who have heard only the wood-thrush may well place him first on the list. He is truly a royal minstrel, and considering his liberal distribution throughout our Atlantic seaboard, perhaps contributes more than any other bird to our sylvan melody. One may object that he spends a little too much time in tuning his instrument, yet his careless and uncertain touches reveal its rare compass and power.

He is the only songster of my acquaintance, excepting the canary, that displays different degrees of proficiency in the exercise of his musical gifts. Not long since, while walking one Sunday in the edge of an orchard adjoining a wood, I heard one that so obviously and unmistakably surpassed all his rivals, that my companion, though slow to notice such things, remarked it wonderingly; and with one accord we paused to listen to so rare a performer. It was not different in quality so much as in quantity. Such a flood of it! Such copiousness! Such long, trilling, accelerating preludes! Such sudden, ecstatic overtures would have intoxicated the dullest ear. He was really without a compeer— a master-artist. Twice afterward I was conscious of having heard the same bird.

The wood-thrush is the handsomest species of this family. In grace and elegance of manner he has no equal. Such a gentle, high-bred air, and such inimitable ease and composure in his flight and movement! He is a poet in very word and deed. His carriage is music to the eye. His performance of the commonest act, as catching a beetle, or picking a worm from the mud, pleases like a stroke of wit or eloquence. Was he a prince in the olden time, and do the regal grace and mien still adhere to him in his transformation? What a finely proportioned form! How plain, yet rich his color,—the bright russet of his back, the clear white of his breast, with the distinct heart-shaped spots! It may be objected to Robin that he is noisy and demonstrative; he hurries away or rises to a branch with an angry note, and flirts his wings in ill-bred suspicion. The mavis, or red-thrush, sneaks and skulks like a culprit, hiding in the densest alders; the cat-bird is a coquette and a flirt, as well as a sort of female Paul Pry; and the chewink shows his inhospitality by espying your movements like a Japanese. The wood-thrush has none of these under-bred traits. He regards me unsuspiciously, or avoids me with a noble reserve,—or, if I am quiet and incurious, graciously hops toward me, as if to pay his respects, or to make my acquaintance. I have passed under his nest within a few feet of his mate and brood, when he sat near by on a branch eying me sharply, but without opening his beak; but the moment I raised my hand toward his defenceless household his anger and indignation were beautiful to behold.

What a noble pride he has! Late one October, after his mates and companions had long since gone South, I noticed one for several successive days in the dense part of this next-door wood, flitting noiselessly about, very grave and silent, as if doing penance for some violation of the code of honor. By many gentle, indirect approaches, I perceived that part of his tail-feathers were undeveloped. The sylvan prince could not think of returning to court in this plight, and so, amid the falling leaves and cold rains of autumn, was patiently biding his time.

The soft, mellow flute of the veery fills a place in the chorus of the woods that the song of the vesper-sparrow fills in the chorus of the fields. It has the nightingale's habit of singing in the twilight, as indeed have all our thrushes. Walk out toward the forest in the warm twilight of a June day, and when fifty rods distant you will hear their soft, reverberating notes, rising from a dozen different throats.

It is one of the simplest strains to be heard,—as simple as the curve in form, delighting from the pure element of harmony and beauty it contains, and not from any novel or fantastic modulation of it,—thus contrasting strongly with such rollicking, hilarious songsters as the bobolink, in whom we are chiefly pleased with the tintinnabulation, the verbal and labial excellence, and the evident conceit and delight of the performer.

I hardly know whether I am more pleased or annoyed with the cat-bird. Perhaps she is a little too common, and her part in the general chorus a little too conspicuous. If you are listening for the note of another bird, she is sure to be prompted to the most loud and protracted singing, drowning all other sounds; if you sit quietly down to observe a favorite or study a new-comer, her curiosity knows no bounds, and you are scanned and ridiculed from every point of observation. Yet I would not miss her; I would only subordinate her a little, make her less conspicuous.

She is the parodist of the woods, and there is ever a mischievous, bantering, half-ironical undertone in her lay, as if she were conscious of mimicking and disconcerting some envied songster. Ambitious of song, practising and rehearsing in private, she yet seems the least sincere and genuine of the sylvan minstrels, as if she had taken up music only to be in the fashion, or not to be outdone by the robins and thrushes. In other words, she seems to sing from some outward motive, and not from inward joyousness. She is a good versifier, but not a great poet. Vigorous, rapid, copious, not without fine touches, but destitute of any high, serene melody, her performance, like that of Thoreau's squirrel, always implies a spectator.

There is a certain air and polish about her strain, however, like that in the vivacious conversation of a well-bred lady of the world, that commands respect. Her maternal instinct, also, is very strong, and that simple structure of dead twigs and dry grass is the centre of much anxious solicitude. Not long since, while strolling through the woods, my attention was attracted to a small densely grown swamp, hedged in with eglantine, brambles, and the everlasting smilax, from which proceeded loud cries of distress and alarm, indicating that some terrible calamity was threatening my sombre-colored minstrel. On effecting an entrance, which, however, was not accomplished till I had doffed coat and hat, so as to diminish the surface exposed to the thorns and brambles, and looking around me from a square yard of terra firma, I found myself the spectator of a loathsome, yet fascinating scene. Three or four yards from me was the nest, beneath which, in long festoons, rested a huge black snake; a bird, two thirds grown, was slowly disappearing between his expanded jaws. As he seemed unconscious of my presence, I quietly observed the proceedings. By slow degrees he compassed the bird about with his elastic mouth; his head flattened, his neck writhed and swelled, and two or three undulatory movements of his glistening body finished the work. Then, he cautiously raised himself up, his tongue flaming from his mouth the while, curved over the nest, and, with wavy, subtle motions, explored the interior. I can conceive of nothing more overpoweringly terrible to an unsuspecting family of birds than the sudden appearance above their domicile of the head and neck of this arch-enemy. It is enough to petrify the blood in their veins. Not finding the object of his search, he came streaming down from the nest to a lower limb, and commenced extending his researches in other directions, sliding stealthily through the branches, bent on capturing one of the parent birds. That a legless, wingless creature should move with such ease and rapidity where only birds and squirrels are considered at home, lifting himself up, letting himself down, running out on the yielding boughs, and traversing with marvellous celerity the whole length and breadth of the thicket, was truly surprising. One thinks of the great myth, of the Tempter and the "cause of all our woe," and wonders if the Arch One is not now playing off some of his pranks before him. Whether we call it snake or devil matters little. I could but admire his terrible beauty, however; his black, shining folds, his easy, gliding movement, head erect, eyes glistening, tongue playing like subtle flame, and the invisible means of his almost winged locomotion.

The parent birds, in the mean while, kept up the most agonizing cry,—at times fluttering furiously about their pursuer, and actually laying hold of his tail with their beaks and claws. On being thus attacked, the snake would suddenly double upon himself and follow his own body back, thus executing a strategic movement that at first seemed almost to paralyze his victim and place her within his grasp. Not quite, however. Before his jaws could close upon the coveted prize the bird would tear herself away, and, apparently faint and sobbing, retire to a higher branch. His reputed powers of fascination availed him little, though it is possible that a frailer and less combative bird might have been held by the fatal spell. Presently, as he came gliding down the slender body of a leaning alder, his attention was attracted by a slight movement of my arm; eying me an instant, with the crouching, utter, motionless gaze which I believe only snakes and devils can assume, he turned quickly,—a feat which necessitated something like crawling over his own body,—and glided off through the branches, evidently recognizing in me a representative of the ancient parties he once so cunningly ruined. A few moments after, as he lay carelessly disposed in the top of a rank alder, trying to look as much like a crooked branch as his supple, shining form would admit, the old vengeance overtook him. I exercised my prerogative, and a well-directed missile, in the shape of a stone, brought him looping and writhing to the ground. After I had completed his downfall and quiet had been partially restored, a half-fledged member of the bereaved household came out from his hiding-place, and, jumping upon a decayed branch, chirped vigorously, no doubt in celebration of the victory.

Till the middle of July there is a general equilibrium; the tide stands poised; the holiday-spirit is unabated. But as the harvest ripens beneath the long, hot days, the melody gradually ceases. The young are out of the nest and must be cared for, and the moulting season is at hand. After the cricket has commenced to drone his monotonous refrain beneath your window, you will not, till another season, hear the wood-thrush in all his matchless eloquence. The bobolink has become careworn and fretful, and blurts out snatches of his song between his scolding and upbraiding, as you approach the vicinity of his nest, oscillating between anxiety for his brood and solicitude for his musical reputation. Some of the sparrows still sing, and occasionally across the hot fields, from a tall tree in the edge of the forest, comes the rich note of the scarlet tanager. This tropical-colored bird loves the hottest weather, and I hear him even in dog-days.

The remainder of the summer is the carnival of the swallows and fly-catchers. Flies and insects, to any amount, are to be had for the catching; and the opportunity is well improved. See that sombre, ashen-colored pewee on yonder branch. A true sportsman, he, who never takes his game at rest, but always on the wing. You vagrant fly, you purblind moth, beware how you come within his range! Observe his attitude, the curious movement of his head, his "eye in a fine frenzy rolling, glancing from heaven to earth, from earth to heaven."

His sight is microscopic and his aim sure. Quick as thought he has seized his victim and is back to his perch. There is no strife, no pursuit, —one fell swoop and the matter is ended. That little sparrow, as you will observe, is less skilled. It is the *Socialis*, and he finds his subsistence properly in various seeds and the larvae of insects, though he occasionally has higher aspirations, and seeks to emulate the pewee, commencing and ending his career as a fly-catcher by an awkward chase after a beetle or "miller." He is hunting around in the grass now, I suspect, with the desire to indulge this favorite whim. There!—the opportunity is afforded him. Away goes a little cream-colored meadow-moth in the most tortuous course he is capable of, and away goes *Socialis* in pursuit. The contest is quite comical, though I dare say it is serious enough to the moth. The chase continues for a few yards, when there is a sudden rushing to cover in the grass,—then a taking to wing again, when the search has become too close, and the moth has recovered his wind. *Socialis* chirps angrily, and is determined not to be beaten. Keeping, with the slightest effort, upon the heels of the fugitive, he is ever on the point of halting to snap him up, but never quite does it,—and so, between disappointment and expectation, is soon disgusted, and returns to pursue his more legitimate means of subsistence.

In striking contrast to this serio-comic strife of the sparrow and the moth is the pigeon-hawk's pursuit of the sparrow or the goldfinch. It is a race of surprising speed and agility. It is a test of wing and wind. Every muscle is taxed, and every nerve strained. Such cries of terror and consternation on the part of the bird, tacking to the right and left, and making the most desperate efforts to escape, and such silent determination on the part of the hawk, pressing the bird so closely, flashing and turning and timing his movements with those of the pursued as accurately and as inexorably as if the two constituted one body, excite feelings of the deepest concern. You mount the fence or rush out of your way to see the issue. The only salvation for the bird is to adopt the tactics of the moth, seeking instantly the cover of some tree, bush, or hedge, where its smaller size enables it to move about more rapidly. These pirates are aware of this, and therefore prefer to take their prey by one fell swoop. You may see one of them prowling through an orchard, with the yellow-birds hovering about him, crying, *Pi-ty*, *pi-ty*, in the most desponding tone; yet he seems not to regard them, knowing, as do they, that in the close branches they are as safe as if in a wall of adamant.

August is the month of the high-sailing hawks. The hen-hawk is the most noticeable. He likes the haze and calm of these long, warm days. He is a bird of leisure, and seems always at his ease. How beautiful and majestic are his movements! So self-poised and easy, such an entire absence of haste, such a magnificent amplitude of circles and spirals, such a haughty, imperial grace, and, occasionally, such daring aerial evolutions!

With slow, leisurely movement, rarely vibrating his pinions, he mounts and mounts in an ascending spiral till he appears a mere speck against the summer sky; then, if the mood seizes him, with wings half closed like a bent bow, he will cleave the air almost perpendicularly, as if intent on dashing himself to pieces against the earth; but, on nearing the ground, he suddenly mounts again on broad, expanded wing, as if rebounding upon the air, and sails leisurely away. It is the sublimest feat of the season. One holds his breath till he sees him rise again.

If inclined to a more gradual and less precipitous descent, he fixes his eye on some distant point in the earth beneath him, and thither bends his course. He is still almost meteoric in his speed and boldness. You see his path down the heavens, straight as a line; if near, you hear the rush of his wings; his shadow hurtles across the fields, and in an instant you see him quietly perched upon some low tree or decayed stub in a swamp or meadow, with reminiscences of frogs and mice stirring in his maw.

When the south wind blows, it is a study to see three or four of these air-kings at the head of the valley far up toward the mountain, balancing and oscillating upon the strong current: now quite stationary, except a slight tremulous motion like the poise of a rope-dancer, then rising and falling in long undulations, and seeming to resign themselves passively to the wind; or, again, sailing high and level far above the mountain's peak, no bluster and haste, but, as stated, occasionally a terrible earnestness and speed. Fire at one as he sails overhead, and, unless wounded badly, he will not change his course or gait.

His flight is a perfect picture of repose in motion. It strikes the eye as more surprising than the flight of the pigeon and swallow even, in that the effort put forth is so uniform and delicate as to escape observation, giving to the movement an air of buoyancy and perpetuity, the effluence of power rather than the conscious application of it.

The calmness and dignity of this hawk when attacked by crows or the king-bird are well worthy of him. He seldom deigns to notice his noisy and furious antagonists, but deliberately wheels about in that aerial spiral, and mounts and mounts till his pursuers grow dizzy and return to earth again. It is quite original, this mode of getting rid of an unworthy opponent, rising to heights where the braggart is dazed and bewildered, and loses his reckoning! I am not sure but it is worthy of imitation.

But summer wanes and autumn approaches. The songsters of the seed-time are silent at the reaping of the harvest. Other minstrels take up the strain. It is the heyday of insect life. The day is canopied with musical sound. All the songs of the spring and summer appear to be floating, softened and refined, in the upper air. The birds in a new, but less holiday suit, turn their faces southward. The swallows flock and go; the bobolinks flock and go; silently and unobserved, the thrushes go. Autumn arrives, bringing finches, warblers, sparrows, and kinglets from the North. Silently the procession passes. Yonder hawk, sailing peacefully away till he is lost in the horizon, is a symbol of the closing season and the departing birds.

WHAT IS A DESERT ? *

EDMUND C. JAEGER

Professor Jaeger has explored the western deserts for the past
forty years. His first journeys into the deserts were on a burro.
Many times he passed slowly through Palm Springs, California,
which was then not yet a busy resort, but still a place of warm,
quiet beauty.

Like John Muir, John Burroughs, and Henry David
Thoreau, Edmund Jaeger accurately records what he sees, feels,
hears, and smells, while studying his beloved deserts. His writ-
ing style is highly entertaining, but also warmly readable.
Professor Jaeger's books include "Desert Wild Flowers," "The
California Deserts," and "Denizens of the Deserts."

* Reprinted from "The North American Deserts" by Edmund C.
Jaeger, with the permission of the publishers, Stanford University Press. Copyright © 1957 by the Board of Trustees of
the Leland Stanford Junior University.

Nearly one-fifth of the surface of the earth is made up of deserts, supporting less than four per cent of the world's population. Although individual parts of these arid regions are quite different in physical appearance, they possess in common several characteristics, such as low rainfall, high average temperatures during the day, and almost constant winds, with consequent increased rate of evaporation.

The most important factor in the creation of a desert is a low annual rainfall. Most geographers have arbitrarily agreed that if a region receives less than ten inches of unevenly distributed rain throughout the year, then it may be termed a desert. The area in question must have in addition a relatively high mean yearly temperature. It is obvious that it would be impossible for a true desert to exist in very cold climates where most of the moisture that falls is retained by freezing and never actually lost. Cold barren regions such as are found in the Arctic and Antarctic can be called wastelands, but certainly not deserts in the true sense.

Because of the high average temperature there is rather rapid evaporation of the little rain that does fall, so that during any year only a small amount of moisture is available for the animals and plants. Naturally the time of the year in which the normal rains occur is important, for if these are mostly in summer, when water loss through evaporation is great, that particular area will be much more arid than one which receives equal amounts of rain in cooler weather.

Much of the summer rain of deserts may be of the cloudburst type, with perhaps several inches falling in a few hours. Most of this moisture is lost to the plants because of the rapid surface run-off. Such rains may cause destructive sheet floods, which are responsible for the burying of young plants under sand and the undercutting of root systems of both immature and mature ones.

Desert areas usually have winds almost constantly sweeping across them, winds that dry out both soil and vegetation. The prevailing winds generally blow *into* the desert from its fringes. The heavy cold air moving into the area replaces the light heated air rising from the desert floor. If the incoming wind is channeled through mountain passes it may be a very steady one and often of considerable force. Such a wind both dries out the country and materially aids the rains in the shifting of soil, causing erosion. The amount of material annually transported by desert winds can be very great.

Almost all deserts are considerably lower in elevation than their surrounding mountains. The streams emptying into them are few. Because most deserts are basins without outlets, the water that collects in their lowest parts soon becomes quite alkaline, and finally, through rapid evaporation, disappears, leaving behind dry lake beds or "clay pans," some of which become heavily encrusted with salts.

Since the plant cover of deserts is necessarily sparse and the amount of solar radiation from the surface soil is great, little of the diurnal heat is retained after the sun goes down. This accounts for the cool, if not cold, nights that often follow very warm days. Because extremes in temperature between day and night are often remarkably high, as much as 70° or 80° F. within a few hours, the animals and plants must be specially adapted to withstand such rapid fluctuations.

The chief causes of deserts are mountains which act as barriers cutting off the moisture-laden clouds that might otherwise sweep across and deposit rain upon them. As moisture-filled clouds blow in and rise they become chilled, and now, no longer able to retain their load of moisture, drop it as rain on the windward side of the mountains away from the desert.

In a few cases, such as in the Atacama Desert of coastal Peru and the Kalahari and Karoo deserts of South Africa, a cold ocean current acts much as mountains do in robbing the rain clouds of moisture.

Each of the continental land masses has its desert. By far the largest in extent is the great Eurasian Palearctic Desert, which includes the Sahara and the deserts of Asia Minor, India, Tibet, China, and Mongolia. Most of the interior of Australia is desert, local names of its parts being the Arunta, Gibson, and Great Victoria deserts. Another major desert area includes much of the southwestern portion of the United States and north-central and northwestern Mexico. For convenience in description, this American Desert region may be divided into five sub-areas: The CHIHUAHUAN, SONORAN, NAVAHOAN, MOHAVEAN, and GREAT BASIN deserts. Of these the Sonoran is sufficiently diversified to warrant a further cleavage into six subdivisions: the SONORA PROPER, ARIZONA UPLAND, YUMAN, COLORADO, VIZCAINO-MAGDALENA, and GULF COAST deserts.

Plants which have become specially adapted to desert conditions are called xerophytes, a word derived from Greek words meaning "dry plants." In order to be able to withstand great heat and severe drought, often over long periods lasting months and occasionally even several years, xerophytic plants have adjusted themselves to their environment in several remarkable ways. They may survive the hot summers by then remaining in the seed stage. They sprout forth only during the rainy season, then grow rapidly, flower, and go back to the seed stage with the advent of the dry season. Most of the small desert annuals are of this type. Their seeds may, if necessary, lie dormant, while buried in the soil, for many years, waiting until a propitious time to begin growth.

Desert plants can also survive by passively evading the dry months. This they do not only by storing and gradually utilizing the moisture received during the year but also by maintaining throughout their lifetime a dwarf form. Such plants are usually widely spaced. Because of their small size they do not need much water to begin with and seem to get along quite well, even to flourish, with unbelievably small amounts of it. Some of the dwarf fragile-stemmed wild buckwheats *(Eriogonum)* are familiar examples.

During dry spells plants may go into a state of dormancy, suspending all normal activity. Such a state is called "drought endurance" by the plant physiologists. Only the advent of rain will arouse such plants to activity. Once the rains come they often leaf out within a period of days. Later the leaves may be almost as quickly shed and there is a sudden return to dormancy. Ocotillos, elephant trees, and jatrophas are desert plants subject to such fluctuations of activity.

Another adaptation is found in those plants which actively resist aridity by storing water within their leaves and stems or roots. They are able to continue growing even through the hottest months when all other desert vegetation suspends almost all growth activity. Many such plants attain to larger than usual bulk and stature and at the same time develop long lateral roots for quickly utilizing surface moisture and long tap roots to reach deep sources of soil water. The various kinds of cacti are examples of such water storers and conservers. Especially typical are the sahuaro and the barrel cacti. During the rainy season the columnar fluted stems become almost "bloated" with water. As the season gets progressively drier and the stored water is gradually used up, the accordion-like stems shrink and become very lean.

When leaves are present on desert plants they are usually greatly modified for water conservation. It is through the leaf that most of the plant's moisture is lost by evaporation. During the warmest part of the day xerophytes often turn or twist their leaves so that only the thin edges are exposed to the sun's direct rays. Leaves that cannot be turned or twisted may curl or roll up during the hotter part of the day and then uncurl in the cooler hours of later afternoon and early morning.

Many desert plants have exceedingly hairy stems and leaves which quickly catch and retain moisture of the surrounding air; the sand verbena is an example. The same hairs may shield the stem and leaf surfaces from direct sun exposure. Leaves of some desert plants have a thick, leathery epidermis to protect them against too rapid water loss; still others, such as the creosote bush and varnish-leaf acacia, have a shiny waxy coating which reflects heat.

Even the breathing pores (stomata) of the leaves of xerophytes exhibit special adaptations. Although many in number, they are typically quite small and either sunken in hair-lined cavities for added protection or shielded by waxy secretions. Further, they may be equipped with valves which close during the day. Normally these vital breathing pores are located on the underside of the leaf away from the sun.

A number of desert plants, such as the smoke tree, crucifixion thorn, and the cacti, have given up leaves almost entirely and have modified their stems to take over leaf functions. In these the outermost coating of the stem, the cuticle, has become toughened and thickened, not only as a means of conserving water but also as a shield against injury and the etching action of wind-borne sand.

A surprisingly large group of xerophytic plants have developed sharp spines or stiffish hairs of one sort or other. Most cacti have them, as do also the acacias, mesquites, condalias, ocotillos, yuccas, and smoke trees. Just why these particular plants have developed a thorny armor is somewhat conjectural, but thorns certainly do, in most cases, act as a protection against being eaten. In addition to spines, some desert plants have developed pungent-odored, bad-tasting, or poisonous substances which deter hungry animals from eating them.

In adapting themselves to severe conditions it is ordinarily the stems and leaves of xerophytes that have changed most of all. The roots have suffered little structural change but they may have altered their manner of growth and distribution. Most of the desert plants have root systems consisting of numerous laterals well spread out and growing close to the surface in order to take quick advantage of the shallow-penetrating rains. A few xerophytes are more or less independent of surface water because of the development of long tap roots reaching to deeper sources of soil water. This is especially true of species which grow on sand dunes such as the mesquite, whose roots may reach well over thirty feet beneath the surface. Roots, both lateral and tap, give firm anchorage against the action of strong winds; the numerous laterals protect against removal of soil from around the base of the plant.

The seemingly top-heavy tree yuccas maintain their upright position, in a land often visited by almost gale-strength winds, by having a resilient trunk and hundreds of long, pencil-sized anchoring rootlets striking almost directly outward from the fringes of their bulging bases. This is also true of the desert palm. Other desert plants such as the sand-mat or rattlesnake weed *(Euphorbia)* protect themselves from the almost constant wind by hugging close to the sand. Many fragile, weak-stemmed annuals grow up through the twiggy stems of rigid shrubs and gain protection against both wind and grazing animals.

Two main problems solved by desert plants also must be met by animals invading a desert environment: getting and preserving vital moisture; and securing adequate protection against excessive heat, sand and dust storms, cool to cold nights, and a host of special enemies.

Unlike plants, which must "stay put" and adapt themselves to weather changes or perish, animals can move about, go underground, or migrate from areas of poor food or water supply to more favorable spots. Thus animals are much more independent of their environment and can to a much greater extent "choose" their preferred habitat.

To conserve water, many animal dwellers venture forth to feed only in the cool of the night. This helps to explain why the desert appears so lifeless to the casual observer who goes abroad only during the day. Animals that forage over considerable reaches of wasteland usually rely for water upon the few scattered springs which they visit at dusk or during the night. Some of the smaller animals that seldom range farther than a few hundred yards from their home secure their water from the plants or animals upon which they feed. Several of the desert rodents, such as the desert hare and kangaroo rat, and many of the wild mice are capable of manufacturing water from their dry food. Such "metabolic water" may enable them to go through life without taking a drink. As a means of further conserving water some desert creatures such as the lizards and snakes void no liquid waste. Both feces and urine are voided in almost solid form.

Most of the desert animals spend much of their period of inactivity beneath the ground, where it is considerably cooler and somewhat moist throughout the year. The heat of the desert sun, even in mid-summer, rarely penetrates more than a few inches below the soil surface, so that underground burrows are many degrees cooler than the air above them. Animals which excavate underground tunnels, such as the kangaroo rat, antelope ground squirrel, pack rat, and shovel-nosed snake, are called fossorial animals. Even the clumsy-appearing desert tortoise digs extensive subterranean chambers and burrows, both for places of hibernation and for shelter from the summer's fierce sun.

Several desert dwellers hibernate or go into a temporary torpor underground during the colder winter months. There is then both an inadequate food supply and a temperature so low that it keeps their bodies from normal functioning. This is especially true of the snakes and most of the lizards. In the case of the ground squirrels and bats (which resort to caves and rock crevices) and that peculiar hibernating bird, the poorwill, lack of food is probably the deciding cause of winter torpidity.

Underground retreats also afford protection from the sand and fine dust raised by the frequent windstorms. As a further protection against sand and dust, Nature has provided many of the desert creatures with smaller than usual ear openings, with the added protection of long hairs or scales partially covering them. The eyelashes may be longer than is normal, and the eyelids thickened. Even the nostrils in some species are provided with valves or can otherwise be tightly closed against sand and dust carried by gusts of wind.

Nearly all desert inhabitants are much lighter colored than their near relatives living in moister climates. This has been cited as an example of a kind of "protective coloration" which enables these creatures to be overlooked by predators. It is assumed that this light color has been evolved over a long period of time. But just how their color change came about is still a problem to be solved by biologists. It is possible, they say, that a paler-colored body reflects more heat and aids in conserving moisture. Only incidentally perhaps does it camouflage the animal against detection by its enemies.

Essentially the same problems that face wild animals had to be solved by the various Indian tribes that made their homes in deserts. The majority of these peoples probably did not come originally into the desert expecting to make it their final home, but were pushed into it by stronger or more warlike tribes on the fringes of the wasteland. However, once they had made an adjustment to the new and severe conditions, they found the desert not only far from inhospitable but a place offering many advantages, such as a warm healthful climate, a ready supply of dry wood for their fires, and numerous fibers and other materials for the making of clothing and shelters.

The first great problem confronting these primitive peoples was that of water. Most of the Indians moved where they could live near lakes, water holes, springs, or some of the few perennial streams. Some of the tribes became more or less sedentary and were able to practice limited agriculture. As they were able to grow more food, they did not have to spend so much time in hunting and therefore had more leisure. With leisure time, it was possible to develop more elaborate rituals and costumes and to become what we term more "civilized." The Hopi pueblo Indians of mid-Arizona, as also their relatives in New Mexico, perhaps reached the highest development of all Indians of the American deserts. The tribes along the Colorado River, such as the Mohave and Yuma, were in some respects even superior to the Hopis. The Navaho and the Apache, coming later into the area, and changing from purely nomadic hunters and raiders into sedentary farmers, have recently become quite civilized and have been able to develop a comparatively high culture.

The most interesting Indian tribes of all are not the sedentary agriculturalists but the food-gatherers and hunters who braved the heat of the desert and not only survived but in some instances even thrived. Such tribes were the Cahuilla of California and the Pima and the Papago of Arizona. These peoples deserve our special respect and admiration as we learn how they experimented with an unusual number of living things of the desert, and found uses for most of them. Predominantly they were nomadic; they had no permanent communities or villages but lived apart as family units, coming together from time to time only for communal hunts or ceremonies. Because they were ever on the move, migrating from one portion of the desert to another as different food plants came into fruit or seed, they could carry only bare essentials and had little time for ornamentation or decoration. Basket making was perhaps their only truly artistic outlet, and in this art these desert tribes excelled. One is sometimes apt to look down upon such people because they were mere food-gatherers and basket-makers. Instead, these tribes should be admired for their persistence and ingenuity in solving the major problems of shelter and subsistence in a thirsty land. There were of necessity few lazy Indians in those by-gone days.

The Cahuilla Indians of southeastern California solved the water problem, at least in some areas, by digging terraced wells in the sand dunes, thus enabling their women to walk to the water level to fill their ollas. Among the staple foods of the desert Cahuilla was the bean of the mesquite (*Prosopis*) and the seeds of the small annual sage called chia. In the higher elevations they utilized for food the young shoots of the agave or century plant *(Agave deserti)*. These they roasted in stone-lined pits. Farther up the desert mountains they harvested the nuts of the piñon or nut pine *(Pinus monophylla)* and gathered acorns, which to make edible they crushed and leached with warm water to remove the poisonous and bitter tannic acid.

The Pima and Papago Indians of southern Arizona and adjacent Mexico depended much upon the flowers and fruit of the sahuaro cactus *(Cereus giganteus)* as a source of food. They also roasted agave shoots and gathered mesquite pods. In many other ways they developed a material culture similar to that of the Cahuilla.

> They that have power to hurt and will do none,
> That do not the thing they most do show. . . .
>
> Shakespeare, Sonnets

MORALS
AND WEAPONS *

KONRAD Z. LORENZ

"King Solomon's Ring" by Konrad Z. Lorenz is one of the most entertaining books ever written on the behavior of animals.

In Lorenz's writings, everyone can gain new understanding and feeling for animals.

Konrad Lorenz, the father of animal behaviorism, demonstrates throughout the book his love for all animals, man included.

Sir Julian Huxley calls Professor Lorenz "one of the most outstanding naturalists of our day."

It is early one Sunday morning at the beginning of March, when Easter is already in the air, and we are taking a walk in the Vienna forest whose wooded slopes of tall beeches can be equalled in beauty by few and surpassed by none. We approach a forest glade. The tall smooth trunks of the beeches soon give place to the Hornbeam which are clothed from top to bottom with pale green foliage. We now tread slowly and more carefully. Before we break through the last bushes and out of cover on to the free expanse of the meadow, we do what all wild animals and all good naturalists, wild boars, leopards, hunters and zoologists would do under similar circumstances: we reconnoitre, seeking, before we leave our cover, to gain from it the advantage which it can offer alike to hunter and hunted, namely, to see without being seen.

Here, too, this age-old strategy proves beneficial. We do actually see someone who is not yet aware of our presence, as the wind is blowing away from him in our direction: in the middle of the clearing sits a large fat hare. He is sitting with his back to us, making a big V with his ears, and is watching intently something on the opposite edge of the meadow. From this point, a second and equally large hare emerges and with slow, dignified hops, makes his way towards the first one. There follows a measured encounter, not unlike the meeting of two strange dogs. This cautious mutual taking stock soon develops into sparring. The two hares chase each other round, head to tail, in minute circles. This giddy rotating continues for quite a long time. Then suddenly, their pent-up energies burst forth into a battle royal. It is just like the outbreak of war, and happens at the very moment when the long mutual threatening of the hostile parties has forced one to the conclusion that neither dares to make a definite move. Facing each other, the hares rear up on their hind legs and, straining to their full height, drum furiously at each other with their fore pads. Now they clash in flying leaps and, at last, to the accompaniment of squeals and grunts, they discharge a volley of lightning kicks, so rapidly that only a slow motion camera could help us to discern the mechanism of these hostilities. Now, for the time being, they have had enough, and they recommence their circling, this time much faster than before; then follows a fresh, more embittered bout. So engrossed are the two champions, that there is nothing to prevent myself and my little daughter from tiptoeing nearer, although that venture cannot be accomplished in silence. Any normal and sensible hare would have heard us long ago, but this is March and March Hares are mad! The whole boxing match looks so comical that my little daughter, in spite of her iron upbringing in the matter of silence when watching animals, cannot restrain a chuckle. That is too much even for March Hares—two flashes in two different directions and the meadow is empty, while over the battlefield floats a fistful of fluff, light as a thistledown.

It is not only funny, it is almost touching, this duel of the unarmed, this raging fury of the meek in heart. But are these creatures really so meek? Have they really got softer hearts than those of the fierce beasts of prey? If, in a zoo, you ever watched two lions, wolves, or eagles in conflict, then, in all probability, you did not feel like laughing. And yet, these sovereigns come off no worse than the harmless hares. Most people have the habit of judging carnivorous and herbivorous animals by quite inapplicable moral criteria. Even in fairy-tales, animals are portrayed as being a community comparable to that of mankind, as though all species of animals were beings of one and the same family, as human beings are. For this reason, the average person tends to regard the animal that kills animals in the same light as he would the man that kills his own kind. He does not judge the fox that kills a hare by the same standard as the hunter who shoots one for precisely the same reason, but with that severe censure that he would apply to the gamekeeper who made a practice of shooting farmers and frying them for supper! The "wicked" beast of prey is branded as a murderer, although the fox's hunting is quite as legitimate and a great deal more necessary to his existence than is that of the gamekeeper, yet nobody regards the latter's "bag" as his prey, and only one author, whose own standards were indicted by the severest moral criticism, has dared to dub the fox-hunter "the unspeakable in pursuit of the uneatable"! In their dealing with members of their own species, the beasts and birds of prey are far more restrained than many of the "harmless" vegetarians.

Still more harmless than a battle of hares appears the fight between turtle- or ring-doves. The gentle pecking of the frail bill, the light flick of the fragile wing seems, to the uninitiated, more like a caress than an attack. Some time ago I decided to breed a cross between the African blond ring-dove and our own indigenous somewhat frailer turtle-dove, and, with this object, I put a tame, home-reared male turtle-dove and a female ring-dove together in a roomy cage. I did not take their original scrapping seriously. How could these paragons of love and virtue dream of harming one another? I left them in their cage and went to Vienna.

When I returned, the next day, a horrible sight met my eyes. The turtle-dove lay on the floor of the cage; the top of his head and neck, as also the whole length of his back, were not only plucked bare of feathers, but so flayed as to form a single wound dripping with blood. In the middle of this gory surface, like an eagle on his prey, stood the second harbinger of peace. Wearing that dreamy facial expression that so appeals to our sentimental observer, this charming lady pecked mercilessly with her silver bill in the wounds of her prostrated mate. When the latter gathered his last resources in a final effort to escape, she set on him again, struck him to the floor with a light clap of her wing and continued with her slow pitiless work of destruction. Without my interference she would undoubtedly have finished him off, in spite of the fact that she was already so tired that she could hardly keep her eyes open. Only in two other instances have I seen similar horrible lacerations inflicted on their own kind by vertebrates: once, as an observer of the embittered fights of cichlid fishes who sometimes actually skin each other, and again as a field surgeon, in the late war, where the highest of all vertebrates perpetrated mass mutilations on members of his own species. But to return to our "harmless" vegetarians. The battle of the hares which we witnessed in the forest clearing would have ended in quite as horrible a carnage as that of the doves, had it taken place in the confines of a cage where the vanquished could not flee the victor.

If this is the extent of the injuries meted out to their own kind by our gentle doves and hares, how much greater must be the havoc wrought amongst themselves by those beasts to whom nature has relegated the strongest weapons with which to kill their prey? One would certainly think so, were it not that a good naturalist should always check by observation even the most obvious-seeming inferences before he accepts them as truth. Let us examine that symbol of cruelty and voraciousness, the wolf. How do these creatures conduct themselves in their dealings with members of their own species? At Whipsnade, that zoological country paradise, there lives a pack of timber wolves. From the fence of a pine-wood of enviable dimensions we can watch their daily round in an environment not so very far removed from conditions of real freedom. To begin with, we wonder why the antics of the many woolly, fat-pawed whelps have not led them to destruction long ago. The efforts of one ungainly little chap to break into a gallop have landed him in a very different situation from that which he intended. He stumbles and bumps heavily into a wicked-looking old sinner.

Strangely enough, the latter does not seem to notice it, he does not even growl. But now we hear the rumble of battle sounds! They are low, but more ominous than those of a dog-fight. We were watching the whelps and have therefore only become aware of this adult fight now that it is already in full swing.

An enormous old timber wolf and a rather weaker, obviously younger one are the opposing champions and they are moving in circles round each other, exhibiting admirable "footwork". At the same time, the bared fangs flash in such a rapid exchange of snaps that the eye can scarcely follow them. So far, nothing has really happened. The jaws of one wolf close on the gleaming white teeth of the other who is on the alert and wards off the attack. Only the lips have received one or two minor injuries. The younger wolf is gradually being forced backwards. It dawns upon us that the older one is purposely manoeuvring him towards the fence. We wait with breathless anticipation what will happen when he "goes to the wall." Now he strikes the wire netting, stumbles . . . and the old one is upon him. And now the incredible happens, just the opposite of what you would expect. The furious whirling of the grey bodies has come to a sudden standstill. Shoulder to shoulder they stand, pressed against each other in a stiff and strained attitude, both heads now facing in the same direction. Both wolves are growling angrily, the elder in a deep bass, the younger in higher tones, suggestive of the fear that underlies his threat. But notice carefully the position of the two opponents; the older wolf has his muzzle close, very close against the neck of the younger, and the latter holds away his head, offering unprotected to his enemy the bend of his neck, the most vulnerable part of his whole body! Less than an inch from the tensed neck-muscles, where the jugular vein lies immediately beneath the skin, gleam the fangs of his antagonist from beneath the wickedly retracted lips. Whereas, during the thick of the fight, both wolves were intent on keeping only their teeth, the one invulnerable part of the body, in opposition to each other, it now appears that the discomfited fighter proffers intentionally that part of his anatomy to which a bite must assuredly prove fatal. Appearances are notoriously deceptive, but in his case, surprisingly, they are not!

The same scene can be watched any time wherever street-mongrels are to be found. I cited wolves as my first example because they illustrate my point more impressively than the all-too-familiar domestic dog. Two adult male dogs meet in the street. Stiff-legged, with tails erect and hair on end, they pace towards each other. The nearer they approach, the stiffer, higher and more ruffled they appear, their advance becomes slower and slower. Unlike fighting cocks they do not make their encounter head to head, front against front, but make as though to pass each other, only stopping when they stand at last flank to flank, head to tail, in close juxtaposition. Then a strict ceremonial demands that each should sniff the hind regions of the other. Should one of the dogs be overcome with fear at this juncture, down goes his tail between his legs and he jumps with a quick, flexible twist, wheeling at an angle of 180 degrees thus modestly retracting his former offer to be smelt. Should the two dogs remain in an attitude of self-display, carrying their tails as rigid as standards, then the sniffing process may be of a long protracted nature. All may be solved amicably and there is still the chance that first one tail and then the other may begin to wag with small but rapidly increasing beats and then this nerve-racking situation may develop into nothing worse than a cheerful canine romp. Failing this solution the situation becomes more and more tense, noses begin to wrinkle and to turn up with a vile, brutal expression, lips begin to curl, exposing the fangs on the side nearer the opponent. Then the animals scratch the earth angrily with their hind feet, deep growls rise from their chests, and, in the next moment, they fall upon each other with loud piercing yells.

But to return to our wolves, whom we left in a situation of acute tension. This was not a piece of inartistic narrative on my part, since the strained situation may continue for a great length of time which is minutes to the observer, but very probably seems hours to the losing wolf. Every second you expect violence and await with bated breath the moment when the winner's teeth will rip the jugular vein of the loser. But your fears are groundless, for it will not happen. In this particular situation, the victor will definitely not close on his less fortunate rival. You can see that he would like to, but he just cannot! A dog or wolf that offers its neck to its adversary in this way will never be bitten seriously. The other growls and grumbles, snaps with his teeth in the empty air and even carries out, without delivering so much as a bite, the movement of shaking something to death in the empty air. However, this strange inhibition from biting persists only so long as the defeated dog or wolf maintains his attitude of humility. Since the fight is stopped so suddenly by this action, the victor frequently finds himself straddling his vanquished foe in anything but a comfortable position. So to remain, with his muzzle applied to the neck of the "under-dog" soon becomes tedious for the champion, and, seeing that he cannot bite anyway, he soon withdraws. Upon this, the under-dog may hastily attempt to put distance between himself and his superior. But he is not usually successful in this, for, as soon as he abandons his rigid attitude of submission, the other again falls upon him like a thunderbolt and the victim must again freeze into his former posture. It seems as if the victor is only waiting for the moment when the other will relinquish his submissive attitude, thereby enabling him to give vent to his urgent desire to bite. But, luckily for the "under-dog," the top-dog at the close of the fight is overcome by the pressing need to leave his trade-mark on the battle-field, to designate it as his personal property—in other words, he must lift his leg against the nearest upright object. This right-of-possession ceremony is usually taken advantage of by the under-dog to make himself scarce.

By this commonplace observation, we are here, as so often, made conscious of a problem which is actual in our daily life and which confronts us on all sides in the most various forms. Social inhibitions of this kind are not rare, but so frequent that we take them for granted and do not stop to think about them. An old German proverb says that one crow will not peck out the eye of another and for once the proverb is right. A tame crow or raven will no more think of pecking at your eye than he will at that of one of his own kind. Often when Roah, my tame raven, was sitting on my arm, I purposely put my face so near to his bill that my open eye came close to its wickedly curved point. Then Roah did something positively touching. With a nervous, worried movement he withdrew his beak from my eye, just as a father who is shaving will hold back his razor blade from the inquisitive fingers of his tiny daughter. Only in one particular connection did Roah ever approach my eye with his bill during this facial grooming. Many of the higher, social birds and mammals, above all monkeys, will groom the skin of a fellow-member of their species in those parts of his body to which he himself cannot obtain access. In birds, it is particularly the head and the region of the eyes which are dependent on the attentions of a fellow. In my description of the jackdaw, I have already spoken of the gestures with which these birds invite one another to preen their head feathers. When, with half-shut eyes, I held my head sideways towards Roah, just as corvine birds do to each other, he understood this movement in spite of the fact that I have no head feathers to ruffle, and at once began to groom me. While doing so, he never pinched my skin, for the epidermis of birds is delicate and would not stand such rough treatment. With wonderful precision, he submitted every attainable hair to a dry-cleaning process by drawing it separately through his bill. He worked with the same intensive concentration that distinguishes the "lousing" monkey and the operating surgeon. This is not meant as a joke: the social grooming of monkeys, and particularly of anthropoid apes has not the object of catching vermin—these animals usually have none—and is not limited to the cleaning of the skin, but serves also more remarkable operations, for instance the dexterous removal of thorns and even the squeezing-out of small carbuncles.

The manipulations of the dangerous-looking corvine beak round the open eye of a man naturally appear ominous and, of course, I was always receiving warnings from onlookers at this procedure. "You never know—a raven is a raven—" and similar words of wisdom. I used to respond with the paradoxical observation that the warner was for me potentially more dangerous than the raven. It has often happened that people have been shot dead by madmen who have masked their condition with the cunning and pretence typical of such cases. There was always a possibility, though admittedly a very small one, that our kind adviser might be afflicted with such a disease. But a sudden and unpredictable loss of the eye-pecking inhibition in a healthy, mature raven is more unlikely by far than an attack by a well-meaning friend.

Why has the dog the inhibition against biting his fellow's neck? Why has the raven an inhibition against pecking the eye of his friend? Why has the ring-dove no such "insurance" against murder? A really comprehensive answer to these questions is almost impossible. It would certainly involve a *historical* explanation of the process by which these inhibitions have been developed in the course of evolution. There is no doubt that they have arisen side by side with the development of the dangerous weapons of the beast of prey. However, it is perfectly obvious why these inhibitions are necessary to all weapon-bearing animals. Should the raven peck, without compunction, at the eye of his nestmate, his wife or his young, in the same way as he pecks at any other moving and glittering object, there would, by now, be no more ravens in the world. Should a dog or wolf unrestrainedly and unaccountably bite the neck of his pack-mates and actually execute the movement of shaking them to death, then his species also would certainly be exterminated within a short space of time.

The ring-dove does not require such an inhibition since it can only inflict injury to a much lesser degree, while its ability to flee is so well developed that it suffices to protect the bird even against enemies equipped with vastly better weapons. Only under the unnatural conditions of close confinement which deprive the losing dove of the possibility of flight does it become apparent that the ring-dove has no inhibitions which prevent it from injuring or even torturing its own kind. Many other "harmless" herbivores prove themselves just as unscrupulous when they are kept in narrow captivity. One of the most disgusting, ruthless and blood-thirsty murderers is an animal which is generally considered as being second only to the dove in the proverbial gentleness of its nature, namely the roe-deer. The roe-buck is about the most malevolent beast I know and is possessed, into the bargain, of a weapon, its antlers, which it shows mighty little restraint in putting into use. The species can "afford" this lack of control since the fleeing capacity even of the weakest doe is enough to deliver it from the strongest buck. Only in very large paddocks can the roe-buck be kept with females of his own kind. In smaller enclosures, sooner or later he will drive his fellows, females and young ones included, into a corner and gore them to death. The only "insurance against murder" which the roe-deer possesses, is based on the fact that the onslaught of the attacking buck proceeds relatively slowly. He does not rush with lowered head at his adversary as, for example, a ram would do, but he approaches quite slowly, cautiously feeling with his antlers for those of his opponent. Only when the antlers are interlocked and the buck feels firm resistance does he thrust with deadly earnest. According to the statistics given by W. T. Hornaday, the former director of the New York Zoo, tame deer cause yearly more serious accidents than captive lions and tigers, chiefly because an uninitiated person does not recognize the slow approach of the buck as an earnest attack, even when the animal's antlers have come dangerously near. Suddenly there follows, thrust upon thrust, the amazingly strong stabbing movement of the sharp weapon, and you will be lucky if you have time enough to get a good grip on the aggressor's antlers. Now there follows a wrestling-match in which the sweat pours and the hands drip blood, and in which even a very strong man can hardly obtain mastery over the roe-buck unless he succeeds in getting to the side of the beast and bending his neck backwards. Of course, one is ashamed to call for help—until one has the point of an antler in one's body! So take my advice and if a charming, tame roe-buck comes playfully towards you, with a characteristic prancing step and flourishing his antlers gracefully, hit him, with your walking stick, a stone or the bare fist, as hard as you can, on the side of his nose, before he can apply his antlers to your person.

And now, honestly judged: who is really a "good" animal, my friend Roah to whose social inhibitions I could trust the light of my eyes, or the gentle ring-dove that in hours of hard work nearly succeeded in torturing its mate to death? Who is a "wicked" animal, the roe-buck who will slit the bellies even of females and young of his own kind if they are unable to escape him, or the wolf who cannot bite his hated enemy if the latter appeals to his mercy?

Now let us turn our mind to another question. Wherein consists the essence of all the gestures of submission by which a bird or animal of a social species can appeal to the inhibitions of its superior? We have just seen, in the wolf, that the defeated animal actually facilitates his own destruction by offering to the victor those very parts of his body which he was most anxious to shield as long as the battle was raging. All submissive attitudes with which we are so far familiar, in social animals, are based on the same principle: The supplicant always offers to his adversary the most vulnerable part of his body, or, to be more exact, that part *against which every killing attack is inevitably directed!* In most birds, this area is the base of the skull. If one jackdaw wants to show submission to another, he squats back on his hocks, turns away his head, at the same time drawing in his bill to make the nape of his neck bulge, and, leaning towards his superior, seems to invite him to peck at the fatal spot. Seagulls and herons present to their superior the top of their head, stretching their neck forward horizontally, low over the ground, also a position which makes the supplicant particularly defenceless.

With many gallinaceous birds, the fights of the males commonly end by one of the combatants being thrown to the ground, held down and then scalped as in the manner described in the ring-dove. Only one species shows mercy in this case, namely the turkey: and this one only does so in response to a specific submissive gesture which serves to forestall the intent of the attack. If a turkey-cock has had more than his share of the wild and grostesque wrestling-match in which these birds indulge, he lays himself with outstretched neck upon the ground. Whereupon the victor behaves exactly as a wolf or dog in the same situation, that is to say, he evidently *wants* to peck and kick at the prostrated enemy, but simply cannot: he would if he could but he can't! So, still in threatening attitude, he walks round and round his prostrated rival, making tentative passes at him, but leaving him untouched.

This reaction—though certainly propitious for the turkey species—can cause a tragedy if a turkey comes to blows with a peacock, a thing which not infrequently happens in captivity, since these species are closely enough related to "appreciate" respectively their mutual manifestations of virility. In spite of greater strength and weight the turkey nearly always loses the match, for the peacock flies better and has a different fighting technique. While the red-brown American is muscling himself up for the wrestling-match, the blue East-Indian has already flown above him and struck at him with his sharply pointed spurs. The turkey justifiably considers this infringement of his fighting code as unfair and, although he is still in possession of his full strength, he throws in the sponge and lays himself down in the above depicted manner now. And a ghastly thing happens: the peacock does not "understand" this submissive gesture of the turkey, that is to say, it elicits no inhibition of his fighting drives. He pecks and kicks further at the helpless turkey, who, if nobody comes to his rescue, is doomed, for the more pecks and blows he receives, the more certainly are his escape reactions blocked by the psycho-physiological mechanism of the submissive attitude. It does not and cannot occur to him to jump up and run away.

The fact that many birds have developed special "signal organs" for eliciting this type of social inhibition, shows convincingly the blind instinctive nature and the great evolutionary age of these submissive gestures. The young of the water-rail, for example, have a bare red patch at the back of their head which, as they present it meaningly to an older and stronger fellow, takes on a deep red colour. Whether, in higher animals and man, social inhibitions of this kind are equally mechanical, need not for the moment enter into our consideration. Whatever may be the reasons that prevent the dominant individual from injuring the submissive one, whether he is prevented from doing so by a simple and purely mechanical reflex process or by a highly philosophical moral standard, is immaterial to the practical issue. The essential behaviour of the submissive as well as of the dominant partner remains the same: the humbled creature suddenly seems to lose his objections to being injured and removes all obstacles from the path of the killer, and it would seem that the very removal of these outer obstacles raises an insurmountable inner obstruction in the central nervous system of the aggressor.

And what is a human appeal for mercy after all? Is it so very different from what we have just described? The Homeric warrior who wishes to yield and plead mercy, discards helmet and shield, falls on his knees and inclines his head, a set of actions which should make it easier for the enemy to kill, but, in reality, hinders him from doing so. As Shakespeare makes Nestor say of Hector:

> Thou hast hung thy advanced sword i' the air,
> Not letting it decline on the declined.

Even to-day, we have retained many symbols of such submissive attitudes in a number of our gestures of courtesy: bowing, removal of the hat, and presenting arms in military ceremonial. If we are to believe the ancient epics, an appeal to mercy does not seem to have raised an "inner obstruction" which was entirely insurmountable. Homer's heroes were certainly not as soft-hearted as the wolves of Whipsnade! In any case, the poet cites numerous instances where the supplicant was slaughtered with or without compunction. The Norse heroic sagas bring us many examples of similar failures of the submissive gesture and it was not till the era of knight-errantry that it was no longer considered "sporting" to kill a man who begged for mercy. The Christian knight is the first who, for reasons of traditional and religious morals, is as chivalrous as is the wolf from the depth of his natural impulses and inhibitions. What a strange paradox!

Of course, the innate, instinctive, fixed inhibitions that prevent an animal from using his weapons indiscriminately against his own kind are only a functional analogy, at the most a slight foreshadowing, a genealogical predecessor of the social morals of man. The worker in comparative ethology does well to be very careful in applying moral criteria to animal behaviour. But here, I must myself own to harbouring sentimental feelings: I think it a truly magnificent thing that one wolf finds himself unable to bite the proffered neck of the other, but still more so that the other relies upon him for this amazing restraint. Mankind can learn a lesson from this, from the animal that Dante calls "la bestia senza pace." I at least have extracted from it a new and deeper understanding of a wonderful and often misunderstood saying from the Gospel which hitherto had only awakened in me feelings of strong opposition: "And unto him that smiteth thee on the one cheek offer also the other" (St. Luke vi, 26). A wolf has enlightened me: not so that your enemy may strike you again do you turn the other cheek toward him, but to make him unable to do it.

When, in the course of its evolution, a species of animals develops a weapon which may destroy a fellow-member at one blow, then, in order to survive, it must develop, along with the weapon, a social inhibition to prevent a usage which could endanger the existence of the species. Among the predatory animals, there are only a few which lead so solitary a life that they can, in general, forego such restraint. They come together only at the mating season when the sexual impulse outweighs all others, including that of aggression. Such unsociable hermits are the polar bear and the jaguar and, owing to the absence of these social inhibitions, animals of these species, when kept together in Zoos, hold a sorry record for murdering their own kind. The system of special inherited impulses and inhibitions, together with the weapons with which a social species is provided by nature, form a complex which is carefully computed and self-regulating. All living beings have received their weapons through the same process of evolution that moulded their impulses and inhibitions; for the structural plan of the body and the system of behaviour of a species are parts of the same whole.

> If such be Nature's holy plan,
> Have I not reason to lament
> What man has made of man?

Wordsworth is right: there is only one being in possession of weapons which do not grow on his body and of whose working plan, therefore, the instincts of his species know nothing and in the usage of which he has no correspondingly adequate inhibition. That being is man. With unarrested growth his weapons increase in monstrousness, multiplying horribly within a few decades. But innate impulses and inhibitions, like bodily structures, need time for their development, time on a scale in which geologists and astronomers are accustomed to calculate, and not historians. We did not receive our weapons from nature. We made them ourselves, of our own free will. Which is going to be easier for us in the future, the production of the weapons or the engendering of the feeling of responsibility that should go along with them, the inhibitions without which our race must perish by virtue of its own creations? We must build up these inhibitions purposefully for we cannot rely upon our instincts. In November 1935, I concluded an article on "Morals and Weapons of Animals" which appeared in a Viennese journal, with the words, "The day will come when two warring factions will be faced with the possibility of each wiping the other out completely. The day may come when the whole of mankind is divided into two such opposing camps. Shall we then behave like doves or like wolves? The fate of mankind will be settled by the answer to this question." We may well be apprehensive.

MIDDLE FORK
OF THE SALMON *

WILLIAM O. DOUGLAS

Justice William O. Douglas is one of America's most distinguished jurists. He is also a rugged outdoorsman who has ardently and eloquently fought for the cause of conservation.

In his book, "My Wilderness—The Pacific West," Justice Douglas describes his rubber raft ride through roaring rapids to the Middle Fork of the Salmon.

Some of Justice Douglas's other books are "Of Men and Mountains," "West of the Indies," "The Right of the People," "America Challenged," "North from Malaya," "Beyond the High Himalayas," and his now-famous "A Wilderness Bill of Rights."

he sixteen-foot rubber boat floated lazily in midstream. The water was so clear and calm I could see the reflection of a granite cliff that towered almost a mile above us. An otter swam noiselessly near shore. A kingfisher dived from an overhanging branch. Somewhere far up the canyon wall an eagle screeched. All else was quiet. Only by sighting a pine on the canyon wall could I tell we were drifting.

Ralph Smothers, my guide, and I drifted in silence. I was at the stern, sitting on the rounded edge of the rubber boat. I leaned against a three-foot stretch of canvas which Ralph had laced to upright aluminum tubing so as to protect passengers from spray. Ralph stood on the "deck" of the rubber boat—two wooden boxes, holding our groceries, that were placed midships. Ralph held a sweep in each hand—one fore and one aft. Each of these sweeps swung on a metal pivot, laced to the boat. They had three-foot blades that could be used to steer the boat or to propel it.

Soon a side current caught the boat and carried it at a brisk speed along the shore. Ralph, who is slight and wiry, moved the boat with deft side movements of the sweeps back to the center of the river, where we caught the main current. Now we drifted at about three miles an hour.

"Grouse Creek Rapids coming up," Ralph said. "We'll need the life jackets here."

We put them on as we rounded a corner of the canyon. Quickly a place that had been deep in solitude was filled with a roar. The rapids were immediately ahead of us.

Grouse Creek Rapids fall about eight feet in fifty yards. The river at this point passes over a ledge that is twelve feet wide and at one point is as sharp as a razor. Rubber boats have been cut to pieces here, when the pilot kept too far to the right. Ralph hit the slick of the falls near the center of the ledge and the boat slid over gracefully. Then it dived, the stern rising high in the air. The nose hit the bottom of the trough and the boat seemed almost to buckle, with its bow on one side of the trough, its stern on the other. Then it started up the big wave, called the rollback. The boat climbed this wall of water in a flash, and for a split second hung over it in mid-air. Then we nose-dived a second time, and the stern flipped so high it showed blue sky between its bottom and the river. We crashed into the bottom of the trough with a thud that sent gallons of water into spray. Ralph turned to me with a grin as the boat leveled off and bobbed like a cork along the minor riffles at the tail of the rapids. With quick side motions of the sweeps he brought the boat to shore to pick up Bob Sandberg and Mercedes, who had been photographing the run from below.

The Middle Fork of the Salmon River in Central Idaho is a fast, white-water river for all of its 130 miles. During the first forty miles it drops sixty feet to the mile. The rest of the way it drops sixteen feet to the mile. It has eleven main rapids in its 130-mile length, not to mention the many pieces of white water that present no special navigation problem. Each of the eleven rapids has its own special risks. In some the danger is from concealed rocks. In others, the position of rocks often makes it necessary to change the course of the boat in the midst of the white water. In some, the main risk is the rollback wave at the bottom of the big drop-off. In every rapids there is the risk of the boat's turning sidewise and being crushed by the rollback.

Though some boatmen use oars or paddles, sweeps are by far the best guarantee of a safe journey. Yet even sweeps present problems. The blade of the sweep can take only a small "bite" of white water. If it dips too far into white water, no man can hold it. If he tries, he'll be yanked into the river. Ralph saw that happen once on Rubber Rapids, a falls so named for the bouncing one gets in a long series of big waves that stretch out a hundred yards or more. The boat behind Ralph entered Rubber Rapids properly and rode the first few waves easily. In a careless second the helmsman lowered the blade of the rear sweep too far into the rapids. The pull of the water was so strong and so sudden that the pilot, who had a fast hold on the sweep, was pulled out of the boat. That was not all. The sweep rebounded, knocking the passenger on the rear seat into the river too. Neither was drowned. But experiences like that make every Middle Fork guide cautious and careful.

The problem of each rapids is different at the various stages of Summer. In July, the best month to run the river, many ledges and rocks are covered that in August are exposed or close to the surface. A change of a few inches in water level creates new navigation hazards. That is why careful guides usually tie up at the head of major rapids and walk down the shore to study the depth of the slick, the size of the rollback, and the position of submerged rocks.

Greater risks come with high water. When the runoff of snow is at its peak, the Middle Fork rises seven feet or more. Then many ledges and sharp corners that present problems in July are rendered harmless.

High water is dangerous. The torrent that pours through some of the funnels in the river at high water throws up rollbacks that no boat suitable for the Middle Fork can survive. These rollbacks flip a boat or swamp it. The history of the Middle Fork during the period of the run-offs is one of tragedy to boatmen. Prudent men do not run the Middle Fork when it is in flood.

Even in the low-water months of July and August there is one piece of white water that cannot be run. It is Sulphur Creek Rapids, some-times known as Dagger Falls. It drops about twenty-five feet in fifty yards. In that drop it passes over two ledges. For small boats the risk of the sheer drop is forbidding. For larger boats the risk is not so much the drop as the ledges.

"I could clear the first ledge with my sweeps," Ralph says. "But one has to keep his sweeps in the white water to steer. If they ever hit that second ledge, I'd be finished."

Ralph lets his boat down Sulphur Creek Rapids on a 200-foot rope. Those who use smaller rubber boats or flat-bottomed wooden boats portage Sulphur Creek Rapids. Those who use the smaller boats do not attempt to run even Grouse Creek Rapids, but let their boats down on ropes instead. But Ralph runs them all except Sulphur Creek Rapids.

Each rapids has a foaming pool with different hydraulics. A few inches to the right or to the left makes the difference between success and tragedy. There is a thrill in doing it just right—and a satisfaction too.

Grouse Creek Rapids, which I described, was for me too breath-less to enjoy. But as the days passed and I studied the river, I came to understand the engineering problems presented by each piece of white water. It was then a joy to see Ralph, a riverman in the best tradition, perform. He seemed to know exactly where the main thrust of the cur-rent was and where the boat need start its downward course in order to avoid the dangerous obstacles. I often watched him catch the main current inches from its center, and so avoid by a hair a dangerous ledge a dozen feet below. Often when we rode a rollback to the peak, it seemed that the boat would turn sideways. It was a thrill in that split second to feel Ralph catch a piece of spray with his rear sweep and straighten us out.

By the time we reached Granite Falls (sometimes called Dirty Drawers) my confidence in Ralph was unbounded. It was good that it was, for Granite Falls to me was the worst of them. Here the river drops eight feet. One enters the rapids over a wide ledge. The channel quickly narrows to a funnel that leads to a large rock on the right. There the water is deflected to the left for a few yards, where it pours onto another huge rock. The danger is in piling up on that second rock.

We dropped off the ledge with a sudden sucking sound and headed for the first rock. An easy touch of the sweeps fore and aft made us miss that rock by inches. Now we were in the main current, headed for the second rock. A touch of the rear sweep turned the nose to the right. The pitch of the stream raised the left side of the boat so that it tipped at a 45-degree angle. In that fashion we rode the roaring wall of water that poured off the right-hand side of the second rock. Not more than three or four seconds had passed since we entered the falls. But in that brief time Ralph had applied the precise pressure on the sweeps to avoid the two rocks and bring us into the clear.

In Porcupine Rapids we had climbed the main rollback and started down when Ralph was nearly thrown from the boat. "Got too big a bite of the white water with my rear sweep," he told me later. Both sweeps were knocked from his hands and Ralph went sprawling in the boat. He was up in a second and back on the deck. Yet in that second the boat had turned completely around in a trough and seemed doomed by an oncoming rollback.

Ralph Smothers was born to the river. His father, A. N. Smothers, has long been a riverman in Idaho. For years he ran the main Salmon River from Salmon City to Riggins. This river was the one that turned back Lewis and Clark, requiring them to take a tortuous overland route to the Snake River and the Columbia. It was on the main Salmon that Ralph, then ten years old, had his first experience with sweeps. He and his younger brother (killed in an airplane crash) were with their father on one of his runs down the main Salmon. The father fell and broke his ankle. They still had twenty miles of treacherous white water to run. Ten-year-old Ralph took the sweeps and under the guidance of his father, who sat at his feet, brought the boat through safely.

Ralph's first trip on the Middle Fork was also with his father. It was in the late thirties, when people in Idaho were still feeling the effect of the depression. There is some gold in the mountains of the Middle Fork—enough to keep a man in bread and beans. Ralph's father and another man went to look for it and took Ralph along. They searched much of the country to the west of the Middle Fork. They combed Bi

Creek, famous in history for skirmishes between the United States Army and the Sheepeater Indians, a branch of the Shoshones. Discouraged, hey followed Big Creek to its junction with the Middle Fork, where hey built a raft to run the thirty miles of the Middle Fork to the Salmon River. All the food they had left was coffee and rice. All went well until hey hit Porcupine Rapids. There the raft was caught in a maelstrom of white water and broken in two. The passengers were thrown clear and managed to get to shore. They salvaged the pieces of the raft and the rice. Only the coffee was lost. In a few hours they had the raft repaired, and reached the main river in safety.

Ralph can make corn bread over a campfire that is as tasty as any in New Orleans. Though good-humored, he is a taciturn chap who seldom talks about himself. His wife, Rae, fills in the details. She seldom goes with Ralph on his trips, as she must stay with the children. When she goes she always gets wet. "You know," she said with a pleasant drawl, "I really think Ralph knows how to make a wave splash anyone in the boat."

Mercedes, remembering the times the waves had broken over us, looked at Ralph quizzically. Ralph grinned and said, "It would be no fun running the Middle Fork if you didn't get a little wet."

Changing the subject, he said, "You always got to figure that the boat will turn over."

"Did you figure that on our trip?" I asked.

"Sure I did," he answered. "Didn't you see my sleeping bag and air mattress wrapped up in that oilskin? I always keep a little air in my mattress, so the bed will float if we capsize."

One can reach the Middle Fork without too much trouble. There is a dirt road out of Stanley, Idaho, leading to Bear Valley Creek. Boats can be hauled to that point. Or one can shorten the trip by flying over 9000-foot ridges and landing on one of the meadows in the upper stretches of the river, taking his rubber boat in the plane. That is what we did. It took three trips from Salmon City to Indian Creek in a Cessna plane, piloted by Mike Loening, to bring us and all our supplies to the Middle Fork. In four hours the job was finished. We could have packed in with horses. But that would have taken days. And time is precious on the Middle Fork in July, when the water is neither too low nor too high.

Once one reaches the Middle Fork he is in solitude that is profound.

The canyon walls, studded with granite cliffs, rise a mile or more. Sometimes they rise sheer a thousand feet. The upper reaches are carved into great, spacious bowls. This region has an interesting geological history. Some 40 million years ago it had been reduced by erosion to gentle contours. Then came the uplift, marked first by the intrusion of granite and later by basalt. During the glacial period the valleys had not yet been cut deep and narrow like a V. The grinding of the ice was felt only at what now are the higher altitudes—mostly above 6500 feet. These glaciers gouged out deep amphitheatrical recesses that now stand out like big bowls above the narrow stream bed. After the ice receded the narrow stream beds were cut by erosion. The land above the river shaped into great basins and capped by granite crags, has only a few slopes that are heavily wooded. This is mostly open country with scatterings of trees and much sagebrush.

There are blue grouse and bobwhites on the slopes and, closer to the river, some ruffed grouse. High up on the ridges and in the bowls grow the white pine and, somewhat lower, the Engelmann's spruce. Down along the waterway are mostly the tenacious black pine, the dignified yellow pine, and the stately Douglas fir. Clumps of aspen often occupy a ravine. Here, too, is the brilliant fireweed. An occasional cottonwood grows near the water's edge, along with stands of willow. Streaks of mountain alder follow streams down the canyon slopes. Huckleberries grow here; so does the elderberry, snowberry, and the wild rose. There are scatterings of hawthorn along the Middle Fork and here and there a hackberry, a chokeberry, and a cascara shrub. One can find Oregon grape and perhaps a Johnny-jump-up. Up and down the canyon are stands of mountain mahogany that elk and deer like for browse. And in the lower reaches of the river the juniper known as Rocky Mountain Red Cedar grows.

Trails touch the river at some points. But not many people travel them these days. There are two dude ranches along the Middle Fork— McCall's and the Flying B Ranch. Yet they are not greatly used. An occasional prospector's cabin still stands. But none is occupied; and the claims once worked have mostly been acquired by the Forest Service.

The result is a 130-mile stretch of white water in a canyon of a remote wilderness. The water is clear, pure, and cold. There are white sand bars without number, where one can make camp under a yellow pine or Douglas fir. In the lower reaches there are rattlesnakes on the ledges above the river. But the sand bars are clear of insects and snakes. These sand bars are filled with enough driftwood to satisfy generations of campers. And I remember some where mountain mahogany, once pressed down by heavy snow, grows almost parallel with the ground and furnishes a convenient roof for sleeping bags. These sand bars have no sign of civilization on them—not even the tin cans and cardboard boxes which usually mark the impact of man on a wilderness. But their fringes are decorated with an occasional blue gentian or some lupine, or perhaps a larkspur. And miner's lettuce and a monkey flower may be found in a shaded, moist place.

Up and down the Middle Fork there are mineral hot springs where deer and sheep come for salting. Here the Sheepeater Indians used to bathe. And today an ingenious traveler can find himself a hot shower bath.

The Middle Fork is one of the finest fishing streams in America. It has cutthroat trout that run up to three pounds and rainbow that run to two pounds. Occasionally a Dolly Varden is caught, and they have been known to run to five pounds. Steelhead and salmon also run the river, coming hundreds of miles from the Pacific to this remote Idaho stream to perform the ritual of spawning.

The prize of the summer fisherman is cutthroat or rainbow trout, both native to the stream. They are so abundant, a party could not possibly eat what its members catch. Our practice was to throw back everything we caught before four o'clock, even the one- or two-pounders. The fishing is so good that there is a drive on by conservationists to ban salmon eggs, spinners, and all bait from the river. The plan is to make the Middle Fork exclusively a fly-fishing stream. Even the fly brings more to the net than one can eat. We used the fly exclusively, concentrating on the bee and caddis patterns. I often fished the flies dry, quartering the river upstream and taking a long float. But I also caught many fish on the reprieve when the fly was wet. The truth is that these Middle Fork trout will strike almost anything that moves or floats. Their pools are rarely disturbed by man. They have not yet developed the wariness of trout that are hunted all Summer long.

There are interesting caves along the Middle Fork. We saw one that was sixty feet long and twenty feet wide. Another lay under an overhang of a granite cliff. Both had petroglyphs on the rock walls. They were made years ago by the Sheepeaters. Those Indians were never more than 200 strong. They were renegades who escaped to the safety of the Middle Fork and lived in its caves, eating mountain sheep and elk that they killed with bows and arrows and skinned with obsidian knives.

No Sheepeaters are left today. They were defeated by the United States Army in 1879 in one of the most difficult military campaigns we have conducted, and were shipped off to an Indian Reservation. Only the writings on the walls of the caves give any clue that man once lived here.

A few years ago Ralph Smothers spent three weeks in the Middle Fork with a man who was dying of cancer. A friend brought the sick man there. The three of them floated the Middle Fork leisurely, so that the dying man could know the full glory of this world before he passed on. There is, indeed, no finer sanctuary in America. The Middle Fork is substantially the wilderness it was a hundred years ago. Its forests have not been cut. The canyons are so remote and so treacherous there has been precious little grazing by cattle and sheep. The few planes that use its meadows have not altered its character. It abounds in game— deer, elk, bear, bobcat, cougar, coyote, mountain sheep, and mountain goat. There are even moose here; and there are also marten, muskrat, mink, and weasel. There are some fresh tracks on every sand bar.

Most of the game is high among the breaks in Summer, coming down late in the Fall. During the ten days we spent in the majestic canyon we saw none except mountain sheep. These sheep are very nervous to any movement above them. An appearance of man on heights that overlook them creates a panic. As long as man stays below them he can approach quite close. One afternoon we spotted ewes and rams on a bench overlooking the river. There were a couple dozen of them, mostly bedded down, only a few grazing. Ralph steered the raft so that we would pass below them. We skimmed the side of the cliff showing stands of a purple penstemon. The sheep were not more than fifteen feet above us. Yet not a one moved.

Back in Salmon City I talked with W. H. Shaw, Supervisor of the Salmon National Forest, about the Middle Fork. His eyes lighted up as he talked of the plans to make this a real wilderness area.

"It's so rugged that trails are not much use. We put all our fire fighters in by parachute these days," he said.

"How do you get them out?" I asked.

"We instruct them to return to the river, where a boat will pick them up," he answered.

The one who will pick them up will be Ralph Smothers, or one of a half-dozen other men who know its white water.

Back in Washington, D.C., I learned that there are engineering plans on file to put as many as nineteen separate dams along the Middle Fork in order to harness it for hydroelectric power. Those of us who have traveled the Middle Fork think this would be the greatest indignity ever inflicted on a sanctuary. The Middle Fork—one of our finest wilderness areas—must be preserved in perpetuity.

Man and his great dams have frequently done more harm than good. Margaret Hindes put the idea in beautiful verse:

> Gone, desecrated for a dam—
> Pines, stream, and trails
> Burned and bared
> Down to dust.
> Now water fills the hollow,
> Water for power,
> But the bowl of wilderness
> Is broken, forever.

I discussed this matter with Olaus J. Murie. "We pay farmers *not* to produce certain crops," I said. "Why not pay the Army Engineers *not* to build dams?"

Olaus laughed and said, "Good idea." And he went on to add that soon all dams for hydroelectric power will be obsolete.

We are, indeed, on the edge of new breakthroughs that will open up sources of power that will make it unnecessary, and indeed foolhardy, to build more dams across our rivers *to produce power*. Hydrogen fusion, with an energy potential that is astronomical, has not yet been mastered. But it certainly will be. Solar energy, though not yet available by commercial standards, is in the offing. Nuclear fission already exists and promises enormous energy supplies. Science may yet save the sanctuary of the Middle Fork from destruction.

PATTERNS
OF SHORE LIFE*

RACHEL L. CARSON

In Miss Carson's "The Edge of the Sea," she opens the door
to a sometimes forgotten world: the world where the teeming,
tenacious life of the sea touches the land. "Each time I enter
this world," she writes, "I gain some new awareness of its
beauty and deeper meanings."

The early history of life as it is written in the rocks is exceedingly dim and fragmentary, and so it is not possible to say when living things first colonized the shore, nor even to indicate the exact time when life arose. The rocks that were laid down as sediments during the first half of the earth's history, in the Archeozoic era, have since been altered chemically and physically by the pressure of many thousands of feet of superimposed layers and by the intense heat of the deep regions to which they have been confined during much of their existence. Only in a few places, as in eastern Canada, are they exposed and accessible for study, but if these pages of the rock history ever contained any clear record of life, it has long since been obliterated.

The following pages—the rocks of the next several hundred million years, known as the Proterozoic era—are almost as disappointing. There are immense deposits of iron, which may possibly have been laid down with the help of certain algae and bacteria. Other deposits—strange globular masses of calcium carbonate—seem to have been formed by lime-secreting algae. Supposed fossils or faint impressions in these ancient rocks have been tentatively identified as sponges, jellyfish, or hard-shelled creatures with jointed legs called anthropods, but the more skeptical or conservative scientists regard these traces as having an inorganic origin.

Suddenly, following the early pages with their sketchy records, a whole section of the history seems to have been destroyed. Sedimentary rocks representing untold millions of years of pre-Cambrian history have disappeared, having been lost by erosion or possibly, through violent changes in the surface of the earth, brought into a location that now is at the bottom of the deep sea. Because of this loss a seemingly unbridgeable gap in the story of life exists.

The scarcity of fossil records in the early rocks and the loss of whole blocks of sediments may be linked with the chemical nature of the early sea and the atmosphere. Some specialists believe that the pre-Cambrian ocean was deficient in calcium or at least in the conditions that make easily possible the secretion of calcium shells and skeletons. If so, its inhabitants must have been for the most part soft-bodied and so not readily fossilized. A large amount of carbon dioxide in the atmosphere and its relative deficiency in the sea would also have affected the weathering of rock, according to geological theory, so that the sedimentary rocks of pre-Cambrian time must have been repeatedly eroded, washed away, and newly sedimented, with consequent destruction of fossils.

When the record is resumed in the rocks of the Cambrian period, which are about half a billion years old, all the major groups of invertebrate animals (including the principal inhabitants of the shore) suddenly appear, fully formed and flourishing. There are sponges and jellyfish, worms of all sorts, a few simple snail-like mollusks, and arthropods. Algae also are abundant, although no higher plants appear. But the basic plan of each of the large groups of animals and plants that now inhabit the shore had been at least projected in those Cambrian seas, and we may suppose, on good evidence, that the strip between the tide lines 500 million years ago bore at least a general resemblance to the intertidal area of the present stage of earth history.

We may suppose also that for at least the preceding half-billion years those invertebrate groups, so well developed in the Cambrian, had been evolving from simpler forms, although what they looked like we may never know. Possibly the larval stages of some of the species now living may resemble those ancestors whose remains the earth seems to have destroyed or failed to preserve.

During the hundreds of millions of years since the dawn of the Cambrian, sea life has continued to evolve. Subdivisions of the original basic groups have arisen, new species have been created, and many of the early forms have disappeared as evolution has developed others better fitted to meet the demands of their world. A few of the primitive creatures of Cambrian time have representatives today that are little changed from their early ancestors, but these are the exception. The shore, with its difficult and changing conditions, has been a testing ground in which the precise and perfect adaptation to environment is an indispensable condition of survival.

All the life of the shore—the past and the present—by the very fact of its existence there, gives evidence that it has dealt successfully with the realities of its world—the towering physical realities of the sea itself, and the subtle life relationships that bind each living thing to its own community. The patterns of life as created and shaped by these realities intermingle and overlap so that the major design is exceedingly complex.

Whether the bottom of the shallow waters and the intertidal area consists of rocky cliffs and boulders, of broad plains of sand, or of coral reefs and shallows determines the visible pattern of life. A rocky coast, even though it is swept by surf, allows life to exist openly through adaptations for clinging to the firm surfaces provided by the rocks and by other structural provisions for dissipating the force of the waves. The visible evidence of living things is everywhere about—a colorful tapestry of seaweeds, barnacles, mussels, and snails covering the rocks —while more delicate forms find refuge in cracks and crevices or by creeping under boulders. Sand, on the other hand, forms a yielding, shifting substratum of unstable nature, its particles incessantly stirred by the waves, so that few living things can establish or hold a place on its surface or even in its upper layers. All have gone below, and in burrows, tubes, and underground chambers the hidden life of the sands is lived. A coast dominated by coral reefs is necessarily a warm coast, its existence made possible by warm ocean currents establishing the climate in which the coral animals can thrive. The reefs, living or dead, provide a hard surface to which living things may cling. Such a coast is somewhat like one bordered by rocky cliffs, but with differences introduced by smothering layers of chalky sediments. The richly varied tropical fauna of coral coasts has therefore developed special adaptations that set it apart from the life of mineral rock or sand. Because the American Atlantic coast includes examples of all three types of shore, the various patterns of life related to the nature of the coast itself are displayed there with beautiful clarity.

Still other patterns are superimposed on the basic geologic ones. The surf dwellers are different from those who live in quiet waters, even if members of the same species. In a region of strong tides, life exists in successive bands or zones, from the high-water mark to the line of the lowest ebb tides; these zones are obscured where there is little tidal action or on sand beaches where life is driven underground. The currents, modifying temperature and distributing the larval stages of sea creatures, create still another world.

Again the physical facts of the American Atlantic coast are such that the observer of its life has spread before him, almost with the clarity of a well-conceived scientific experiment, a demonstration of the modifying effect of tides, surf, and currents. It happens that the northern rocks, where life is lived openly, lie in the region of some of the strongest tides of the world, those within the area of the Bay of Fundy. Here the zones of life created by the tides have the simple graphic force of a diagram. The tidal zones being obscured on sandy shores, one is free there to observe the effect of the surf. Neither strong tides nor heavy surf visits the southern tip of Florida. Here is a typical coral coast, built by the coral animals and the mangroves that multiply and spread in the calm, warm waters—a world whose inhabitants have drifted there on ocean currents from the West Indies, duplicating the strange tropical fauna of that region.

And over all these patterns there are others created by the sea water itself—bringing or withholding food, carrying substances of powerful chemical nature that, for good or ill, affect the lives of all they touch. Nowhere on the shore is the relation of a creature to its surroundings a matter of a single cause and effect; each living thing is bound to its world by many threads, weaving the intricate design of the fabric of life.

The problem of breaking waves need not be faced by inhabitants of the open ocean, for they can sink into deep water to avoid rough seas. An animal or plant of the shore has no such means of escape. The surf releases all its tremendous energy as it breaks against the shore, sometimes delivering blows of almost incredible violence. Exposed coasts of Great Britain and other eastern Atlantic islands receive some of the most violent surf in the world, created by winds that sweep across the whole expanse of ocean. It sometimes strikes with a force of two tons to the square foot. The American Atlantic coast, being a sheltered shore, receives no such surf, yet even here the waves of winter storms or of summer hurricanes have enormous size and destructive power. The island of Monhegan on the coast of Maine lies unprotected in the path of such storms and receives their waves on its steep seaward-facing cliffs. In a violent storm the spray from breaking waves is thrown over the crest of White Head, about 100 feet above the sea. In some storms the green water of actual waves sweeps over a lower cliff known as Gull Rock. It is about 60 feet high.

The effect of waves is felt on the bottom a considerable distance offshore. Lobster traps set in water nearly 200 feet deep often are shifted about or have stones carried into them. But the critical problem, of course, is the one that exists on or very close to the shore, where waves are breaking. Very few coasts have completely defeated the attempts of living things to gain a foothold. Beaches are apt to be barren if they are composed of loose coarse sand that shifts in the surf and then dries quickly when the tide falls. Others, of firm sand, though they may look barren, actually sustain a rich fauna in their deeper layers. A beach composed of many cobblestones that grind against each other in the surf is an impossible home for most creatures. But the shore formed of rocky cliffs and ledges, unless the surf be of extraordinary force, is host to a large and abundant fauna and flora.

Barnacles are perhaps the best example of successful inhabitants of the surf zone. Limpets do almost as well, and so do the small rock periwinkles. The coarse brown seaweeds called wracks or rockweeds possess species that thrive in moderately heavy surf, while others require a degree of protection. After a little experience one can learn to judge the exposure of any shore merely by identifying its fauna and flora. If, for example, there is a broad area covered by the knotted wrack—a long and slender weed that lies like a tangled mass of cordage when the tide is out—if this predominates, we know the shore is a moderately protected one, seldom visited by heavy surf. If, however, there is little or none of the knotted wrack but instead a zone covered by a rockweed of much shorter stature, branching repeatedly, its fronds flattened and tapering at the ends, then we sense more keenly the presence of the open sea and the crushing power of its surf. For the forked wrack and other members of a community of low-growing seaweeds with strong and elastic tissues are sure indicators of an exposed coast and can thrive in seas the knotted wrack cannot endure. And if, on still another shore, there is little vegetation of any sort, but instead only a rock zone whitened by a living snow of barnacles—thousands upon thousands of them raising their sharp-pointed cones to the smother of the surf—we may be sure this coast is quite unprotected from the force of the sea.

The barnacle has two advantages that allow it to succeed where almost all other life fails to survive. Its low conical shape deflects the force of the waves and sends the water rolling off harmlessly. The whole base of the cone, moreover, is fixed to the rock with natural cement of extraordinary strength; to remove it one has to use a sharp-bladed knife. And so those twin dangers of the surf zone—the threat of being washed away and of being crushed—have little reality for the barnacle. Yet its existence in such a place takes on a touch of the miraculous when we remember this fact: it was not the adult creature, whose shape and firmly cemented base are precise adaptations to the surf, that gained a foothold here; it was the larva. In the turbulence of heavy seas, the delicate larva had to choose its spot on the wave-washed rocks, to settle there, and somehow not be washed away during those critical hours while its tissues were being reorganized in their transformation to the adult form, while the cement was extruded and hardened, and the shell plates grew up about the soft body. To accomplish all this in heavy surf seems to me a far more difficult thing than is required of the spore of a rockweed; yet the fact remains that the barnacles can colonize exposed rocks where the weeds are unable to gain a footing.

The streamlined form has been adopted and even improved upon by other creatures, some of whom have omitted the permanent attachment to the rocks. The limpet is one of these—a simple and primitive snail that wears above its tissues a shell like the hat of a Chinese coolie. From this smoothly sloping cone the surf rolls away harmlessly; indeed, the blows of falling water only press down more firmly the suction cup of fleshy tissue beneath the shell, strengthening its grip on the rock.

Still other creatures, while retaining a smoothly rounded contour, put out anchor lines to hold their places on the rocks. Such a device is used by the mussels, whose numbers in even a limited area may be almost astronomical. The shells of each animal are bound to the rock by a series of tough threads, each of shining silken appearance. The threads are a kind of natural silk, spun by a gland in the foot. These anchor lines extend out in all directions; if some are broken, the others hold while the damaged lines are being replaced. But most of the threads are directed forward and in the pounding of storm surf the mussel tends to swing around and head into the seas, taking them on the narrow "prow" and so minimizing their force.

Even the sea urchins can anchor themselves firmly in moderately strong surf. Their slender tube feet, each equipped with a suction disc at its tip, are thrust out in all directions. I have marveled at the green urchins on a Maine shore, clinging to the exposed rock at low water of spring tides, where the beautiful coralline algae spread a rose-colored crust beneath the shining green of their bodies. At that place the bottom slopes away steeply and when the waves at low tide break on the crest of the slope, they drain back to the sea with a strong rush of water. Yet as each wave recedes, the urchins remain on their accustomed stations, undisturbed.

For the long-stalked kelps that sway in dusky forests just below the level of the spring tides, survival in the surf zone is largely a matter of chemistry. Their tissues contain large amounts of alginic acid and its salts, which create a tensile strength and elasticity able to withstand the pulling and pounding of the waves.

Still others—animal and plant—have been able to invade the surf zone by reducing life to a thin creeping mat of cells. In such form many sponges, ascidians, bryozoans, and algae can endure the force of waves. Once removed from the shaping and conditioning effect of surf, however, the same species may take on entirely different forms. The pale green crumb-of-bread sponge lies flat and almost paper-thin on rocks facing toward the sea; back in one of the deep rock pools its tissues build up into thickened masses, sprinkled with the cone-and-crater structure that is one of the marks of the species. Or the golden-star tunicate may expose a simple sheet of jelly to the waves, though in quiet water it hangs down in pendulous lobes flecked with the starry forms of the creatures that comprise it.

As on the sands almost everything has learned to endure the surf by burrowing down to escape it, so on the rocks some have found safety by boring. Where ancient marl is exposed on the Carolina coast, it is riddled by date mussels. Masses of peat contain the delicately sculptured shells of mollusks called angel wings, seemingly fragile as china, but nevertheless able to bore into clay or rock; concrete piers are drilled by small boring clams; wooden timbers by other clams and isopods. All of these creatures have exchanged their freedom for a sanctuary from the waves, being imprisoned forever within the chambers they have carved.

The vast current systems, which flow through the oceans like rivers, lie for the most part offshore and one might suppose their influence in intertidal matters to be slight. Yet the currents have far-reaching effects, for they transport immense volumes of water over long distances—water that holds its original temperature through thousands of miles of its journey. In this way tropical warmth is carried northward and arctic cold brought far down toward the equator. The currents, probably more than any other single element, are the creators of the marine climate.

The importance of climate lies in the fact that life, even as broadly defined to include all living things of every sort, exists within a relatively narrow range of temperature, roughly between 32° F. and 210° F. The planet Earth is particularly favorable for life because it has a fairly stable temperature. Especially in the sea, temperature changes are moderate and gradual and many animals are so delicately adjusted to the accustomed water climate that an abrupt or drastic change is fatal. Animals living on the shore and exposed to air temperatures at low tide are necessarily a little more hardy, but even these have their preferred range of heat and cold beyond which they seldom stray.

Most tropical animals are more sensitive to change—especially toward higher temperatures—than northern ones, and this is probably because the water in which they live normally varies by only a few degrees throughout the year. Some tropical sea urchins, keyhole limpets, and brittle stars die when the shallow waters heat to about 99° F. The arctic jellyfish Cyanea, on the other hand, is so hardy that it continues to pulsate when half its bell is imprisoned in ice, and may revive even after being solidly frozen for hours. The horseshoe crab is an example of an animal that is very tolerant of temperature change. It has a wide range as a species, and its northern forms can survive being frozen into ice in New England, while its southern representatives thrive in tropical waters of Florida and southward to Yucatán.

Shore animals for the most part endure the seasonal changes of temperate coasts, but some find it necessary to escape the extreme cold of winter. Ghost crabs and beach fleas are believed to dig very deep holes in the sand and go into hibernation. Mole crabs that feed in the surf much of the year retire to the bottom offshore in winter. Many of the hydroids, so like flowering plants in appearance, shrink down to the very core of their animal beings in winter, withdrawing all living tissues into the basal stalk. Other shore animals, like annuals in the plant kingdom, die at the end of summer. All of the white jellyfish, so common in coastal waters during the summer, are dead when the last autumn gale has blown itself out, but the next generation exists as little plant-like beings attached to the rocks below the tide.

For the great majority of shore inhabitants that continue to live in the accustomed places throughout the year, the most dangerous aspect of winter is not cold but ice. In years when much shore ice is formed, the rocks may be scraped clean of barnacles, mussels, and seaweeds simply by the mechanical action of ice grinding in the surf. After this happens, several growing seasons separated by moderate winters may be needed to restore the full community of living creatures.

Because most sea animals have definite preferences as to aquatic climate, it is possible to divide the coastal waters of eastern North America into zones of life. While variation in the temperature of the water within these zones is in part a matter of the advance from southern to northern latitudes, it is also strongly influenced by the pattern of the ocean currents—the sweep of warm tropical water carried northward in the Gulf Stream, and the chill Labrador Current creeping down from the north on the landward border of the Stream, with complex intermixing of warm and cold water between the boundaries of the currents.

From the point where it pours through the Florida straits up as far as Cape Hatteras, the Stream follows the outer edge of the continental shelf, which varies greatly in width. At Jupiter Inlet on the east coast of Florida this shelf is so narrow that one can stand on shore and look out across emerald-green shallows to the place where the water suddenly takes on the intense blue of the Stream. At about this point there seems to exist a temperature barrier, separating the tropical fauna of southern Florida and the Keys from the warm-temperate fauna of the area lying between Cape Kennedy and Cape Hatteras. Again at Hatteras the shelf becomes narrow, the Stream swings closer inshore, and the northward-moving water filters through a confused pattern of shoals and sub-merged sandy hills and valleys. Here again is a boundary between life zones, though it is a shifting and far from absolute one. During the winter, temperatures at Hatteras probably forbid the northward passage of migratory warm-water forms, but in summer the temperature bar-riers break down, the invisible gates open, and these same species may range far toward Cape Cod.

From Hatteras north the shelf broadens, the Stream moves far off-shore, and there is a strong infiltration and mixing of colder water from the north, so that the progressive chilling is speeded. The difference in temperature between Hatteras and Cape Cod is as great as one would find on the opposite side of the Atlantic between the Canary Islands and southern Norway—a distance five times as long. For migratory sea fauna this is an intermediate zone, which cold-water forms enter in winter, and warm-water species in summer. Even the resident fauna has a mixed, indeterminate character, for this area seems to receive some of the more temperature-tolerant forms from both north and south, but to have few species that belong to it exclusively.

Cape Cod has long been recognized in zoology as marking the boundary of the range for thousands of creatures. Thrust far into the sea, it interferes with the passage of the warmer waters from the south and holds the cold waters of the north within the long curve of its shore. It is also a point of transition to a different kind of coast. The long sand strands of the south are replaced by rocks, which come more and more to dominate the coastal scene. They form the sea bottom as well as its shores; the same rugged contours that appear in the land forms of this region lie drowned and hidden from view offshore. Here zones of deep water, with accompanying low temperatures, lie generally closer to the shore than they do farther south, with interesting local effects on the populations of shore animals. Despite the deep inshore waters, the numerous islands and the jaggedly indented coast create a large intertidal area and so provide for a rich shore fauna. This is the cold-temperate region, inhabited by many species unable to tolerate the warm water south of the Cape. Partly because of the low temperatures and partly because of the rocky nature of the shore, heavy growths of seaweeds cover the ebb-tide rocks with a blanket of various hues, herds of periwinkles graze, and the shore is here whitened by millions of barnacles or there darkened by millions of mussels.

Beyond, in the waters bathing Labrador, southern Greenland, and parts of Newfoundland, the temperature of the sea and the nature of its flora and fauna are subarctic. Still farther to the north is the arctic province, with limits not yet precisely defined.

Although these basic zones are still convenient and well-founded divisions of the American coast, it became clear by about the third decade of the twentieth century that Cape Cod was not the absolute barrier it had once been for warm-water species attempting to round it from the south. Curious changes have been taking place, with many animals invading this cold-temperate zone from the south and pushing up through Maine and even into Canada. This new distribution is, of course, related to the widespread change of climate that seems to have set in about the beginning of the century and is now well recognized— a general warming-up noticed first in arctic regions, then in subarctic, and now in the temperate areas of northern states. With warmer ocean waters north of Cape Cod, not only the adults but the critically important young stages of various southern animals have been able to survive.

One of the most impressive examples of northward movement is provided by the green crab, once unknown north of the Cape, now familiar to every clam fisherman in Maine because of its habit of preying on the young stages of the clam. Around the turn of the century, zoological manuals gave its range as New Jersey to Cape Cod. In 1905 it was reported near Portland, and by 1930 specimens had been collected in Hancock County, about midway along the Maine coast. During the following decade it moved along to Winter Harbor, and in 1951 was found at Lubec. Then it spread up along the shores of Passamaquoddy Bay and crossed to Nova Scotia.

With higher water temperatures the sea herring is becoming scarce in Maine. The warmer waters may not be the only cause, but they are undoubtedly responsible in part. As the sea herring decline, other kinds of fish are coming in from the south. The menhaden is a larger member of the herring family, used in enormous quantities for manufacturing fertilizer, oils, and other industrial products. In the 1880's there was a fishery for menhaden in Maine, then they disappeared and for many years were confined almost entirely to areas south of New Jersey. About 1950, however, they began to return to Maine waters, followed by Virginia boats and fishermen. Another fish of the same tribe, called the round herring, is also ranging farther north. In the 1920's Professor Henry Bigelow of Harvard University reported it as occurring from the Gulf of Mexico to Cape Cod, and pointed out that it was rare anywhere on the Cape. (Two caught at Provincetown were preserved in the Museum of Comparative Zoölogy at Harvard.) In the 1950's, however, immense schools of this fish appeared in Maine waters, and the fishing industry began experiments with canning it.

Many other scattered reports follow the same trend. The mantis shrimp, formerly barred by the Cape, has now rounded it and spread into the southern part of the Gulf of Maine. Here and there the soft-shell clam shows signs of being adversely affected by warm summer temperatures and the hard-shell species is replacing it in New York waters. Whiting, once only summer fish north of the Cape, now are caught there throughout the year, and other fish once thought distinctively southern are able to spawn along the coast of New York, where their delicate juvenile stages formerly were killed by the cold winters.

Despite the present exceptions, the Cape Cod–Newfoundland coast is typically a zone of cool waters inhabited by a boreal flora and fauna. It displays strong and fascinating affinities with distant places of the northern world, linked by the unifying force of the sea with arctic waters and with the coasts of the British Isles and Scandinavia. So many of its species are duplicated in the eastern Atlantic that a handbook for the British Isles serves reasonably well for New England, covering probably 80 per cent of the seaweeds and 60 per cent of the marine animals. On the other hand, the American boreal zone has stronger ties with the arctic than does the British coast. One of the large Laminarian seaweeds, the arctic kelp, comes down to the Maine coast but is absent in the eastern Atlantic. An arctic sea anemone occurs in the western North Atlantic abundantly down to Nova Scotia and less numerously in Maine, but on the other side misses Great Britain and is confined to colder waters farther north. The occurrence of many species such as the green sea urchin, the blood-red starfish, the cod, and the herring are examples of a distribution that is circumboreal, extending right around the top of the earth and brought about through the agency of cold currents from melting glaciers and drifting pack ice that carry representatives of the northern faunas down into the North Pacific and North Atlantic.

The existence of so strong a common element between the faunas and floras of the two coasts of the North Atlantic suggests that the means of crossing must be relatively easy. The Gulf Stream carries many migrants away from American shores. The distance to the opposite side is great, however, and the situation is complicated by the short larval life of most species and the fact that shallow waters must be within reach when the time comes for assuming the life of the adult. In this northern part of the Atlantic intermediate way-stations are provided by submerged ridges, shallows, and islands, and the crossing may be broken into easy stages. In some earlier geologic times these shallows were even more extensive, so over long periods both active and involuntary migration across the Atlantic have been feasible.

In lower latitudes the deep basin of the Atlantic must be crossed, where few islands or shallows exist. Even here some transfer of larvae and adults takes place. The Bermuda Islands, after being raised above the sea by volcanic action, received their whole fauna as immigrants from the West Indies via the Gulf Stream. And on a smaller scale the long transatlantic crossings have been accomplished. Considering the difficulties, an impressive number of West Indian species are identical with, or closely related to African species, apparently having crossed in the Equatorial Current. They include species of starfish, shrimp, crayfish, and mollusks. Where such a long crossing has been made it is logical to assume that the migrants were adults, traveling on floating timber or drifting seaweed. In modern times, several African mollusks and starfish have been reported as arriving at the Island of St. Helena by these means.

The records of paleontology provide evidence of the changing shapes of continents and the changing flow of the ocean currents, for these earlier earth patterns account for the otherwise mysterious present distribution of many plants and animals. Once, for example, the West Indian region of the Atlantic was in direct communication, via sea currents, with the distant waters of the Pacific and Indian Oceans. Then a land bridge built up between the Americas, the Equatorial Current turned back on itself to the east, and a barrier to the dispersal of sea creatures was erected. But in species living today we find indications of how it was in the past. Once I discovered a curious little mollusk living in a meadow of turtle grass on the floor of a quiet bay among Florida's Ten Thousand Islands. It was the same bright green as the grass, and its little body was much too large for its thin shell, out of which it bulged. It was one of the scaphanders, and its nearest living relatives are inhabitants of the Indian Ocean. And on the beaches of the Carolinas I have found rocklike masses of calcareous tubes, secreted by colonies of a dark-bodied little worm. It is almost unknown in the Atlantic; again its relatives are Pacific and Indian Ocean forms.

And so transport and wide dispersal are a continuing, universal process—an expression of the need of life to reach out and occupy all habitable parts of the earth. In any age the pattern is set by the shape of the continents and the flow of the currents; but it is never final, never completed.

On a shore where tidal action is strong and the range of the tide is great, one is aware of the ebb and flow of water with a daily, hourly awareness. Each recurrent high tide is a dramatic enactment of the advance of the sea against the continents, pressing up to the very threshold of the land, while the ebbs expose to view a strange and unfamiliar world. Perhaps it is a broad mud flat where curious holes, mounds, or tracks give evidence of a hidden life alien to the land; or perhaps it is a meadow of rockweeds lying prostrate and sodden now that the sea has left them, spreading a protective cloak over all the animal life beneath them. Even more directly the tides address the sense of hearing, speaking a language of their own distinct from the voice of the surf. The sound of a rising tide is heard most clearly on shores removed from the swell of the open ocean. In the stillness of night the strong waveless surge of a rising tide creates a confused tumult of water sounds—swashings and swirlings and a continuous slapping against the rocky rim of the land. Sometimes there are undertones of murmurings and whisperings; then suddenly all lesser sounds are obliterated by a torrential inpouring of water.

On such a shore the tides shape the nature and behavior of life. Their rise and fall give every creature that lives between the high- and low-water lines a twice-daily experience of land life. For those that live near the low-tide line the exposure to sun and air is brief; for those higher on the shore the interval in an alien environment is more prolonged and demands greater powers of endurance. But in all the inter-tidal area the pulse of life is adjusted to the rhythm of the tides. In a world that belongs alternately to sea and land, marine animals, breathing oxygen dissolved in sea water, must find ways of keeping moist; the few air breathers who have crossed the high-tide line from the land must protect themselves from drowning in the flood tide by bringing with them their own supply of oxygen. When the tide is low there is little or no food for most intertidal animals, and indeed the essential processes of life usually have to be carried on while water covers the shore. The tidal rhythm is therefore reflected in a biological rhythm of alternating activity and quiescence.

On a rising tide, animals that live deep in sand come to the surface, or thrust up the long breathing tubes or siphons, or begin to pump water through their burrows. Animals fixed to rocks open their shells or reach out tentacles to feed. Predators and grazers move about actively. When the water ebbs away the sand dwellers withdraw into the deep wet layers; the rock fauna brings into use all its varied means for avoiding desiccation. Worms that build calcareous tubes draw back into them, sealing the entrance with a modified gill filament that fits like a cork in a bottle. Barnacles close their shells, holding the moisture around their gills. Snails draw back into their shells, closing the door-like operculum to shut out the air and keep some of the sea's wetness within. Scuds and beach fleas hide under rocks or weeds, waiting for the incoming tide to release them.

All through the lunar month, as the moon waxes and wanes, so the moon-drawn tides increase or decline in strength and the lines of high and low water shift from day to day. After the full moon, and again after the new moon, the forces acting on the sea to produce the tide are stronger than at any other time during the month. This is because the sun and moon then are directly in line with the earth and their attractive forces are added together. For complex astronomical reasons, the greatest tidal effect is exerted over a period of several days immediately after the full and the new moon, rather than at a time precisely coinciding with these lunar phases. During these periods the flood tides rise higher and the ebb tides fall lower than at any other time. These are called the "spring tides" from the Saxon "sprungen." The word refers not to a season, but to the brimming fullness of the water causing it to "spring" in the sense of a strong, active movement. No one who has watched a new-moon tide pressing against a rocky cliff will doubt the appropriateness of the term. In its quarter phases, the moon exerts its attraction at right angles to the pull of the sun so the two forces interfere with each other and the tidal movements are slack. Then the water neither rises as high nor falls as low as on the spring tides. These sluggish tides are called the "neaps"—a word that goes back to old Scandinavian roots meaning "barely touching" or "hardly enough."

On the Atlantic coast of North America the tides move in the so-called semidiurnal rhythm, with two high and two low waters in each tidal day of about 24 hours and 50 minutes. Each low tide follows the previous low by about 12 hours and 25 minutes, although slight local variations are possible. A like interval, of course, separates the high tides.

The range of tide shows enormous differences over the earth as a whole and even on the Atlantic coast of the United States there are important variations. There is a rise and fall of only a foot or two around the Florida Keys. On the long Atlantic coast of Florida the spring tides have a range of 3 to 4 feet, but a little to the north, among the Sea Islands of Georgia, these tides have an 8-foot rise. Then in the Carolinas and northward to New England they move less strongly, with spring tides of 6 feet at Charleston, South Carolina, 3 feet at Beaufort, North Carolina, and 5 feet at Cape May, New Jersey. Nantucket Island has little tide, but on the shores of Cape Cod Bay, less than 30 miles away, the spring tide range is 10 to 11 feet. Most of the rocky coast of New England falls within the zone of the great tides of the Bay of Fundy. From Cape Cod to Passamaquoddy Bay the amplitude of their range varies but is always considerable: 10 feet at Provincetown, 12 at Bar Harbor, 20 at Eastport, 22 at Calais. The conjunction of strong tides and a rocky shore, where much of the life is exposed, creates in this area a beautiful demonstration of the power of the tides over living things.

As day after day these great tides ebb and flow over the rocky rim of New England, their progress across the shore is visibly marked in stripes of color running parallel to the sea's edge. These bands, or zones, are composed of living things and reflect the stages of the tide, for the length of time that a particular level of shore is uncovered determines, in large measure, what can live there. The hardiest species live in the upper zones. Some of the earth's most ancient plants—the blue-green algae—though originating eons ago in the sea, have emerged from it to form dark tracings on the rocks above the high-tide line, a black zone visible on rocky shores in all parts of the world. Below the black zone, snails that are evolving toward a land existence browse on the film of vegetation or hide in seams and crevices in the rocks. But the most conspicuous zone begins at the upper line of the tides. On an open shore with moderately heavy surf, the rocks are whitened by the crowded millions of the barnacles just below the high-tide line. Here and there the white is interrupted by mussels growing in patches of darkest blue. Below them the seaweeds come in—the brown fields of the rockweeds. Toward the low-tide line the Irish moss spreads its low cushioning growth—a wide band of rich color that is not fully exposed by the sluggish movements of some of the neap tides, but appears on all of the greater tides. Sometimes the reddish brown of the moss is splashed with the bright green tangles of another seaweed, a hairlike growth of wiry texture. The lowest of the spring tides reveal still another zone during the last hour of their fall—that sub-tide world where all the rock is painted a deep rose hue by the lime-secreting seaweeds that encrust it, and where the gleaming brown ribbons of the large kelps lie exposed on the rocks.

With only minor variations, this pattern of life exists in all parts of the world. The differences from place to place are related usually to the force of the surf, and one zone may be largely suppressed and another enormously developed. The barnacle zone, for example, spreads its white sheets over all the upper shore where waves are heavy, and the rockweed zone is greatly reduced. With protection from surf, the rockweeds not only occupy the middle shore in profusion but invade the upper rocks and make conditions difficult for the barnacles.

Perhaps in a sense the true intertidal zone is that band between high and low water of the neap tides, an area that is completely covered and uncovered during each tidal cycle, or twice during every day. Its inhabitants are the typical shore animals and plants, requiring some daily contact with the sea but able to endure limited exposure to land conditions.

Above high water of neaps is a band that seems more of earth than of sea. It is inhabited chiefly by pioneering species; already they have gone far along the road toward land life and can endure separation from the sea for many hours or days. One of the barnacles has colonized these higher high-tide rocks, where the sea comes only a few days and nights out of the month, on the spring tides. When the sea returns it brings food and oxygen, and in season carries away the young into the nursery of the surface waters; during these brief periods the barnacle is able to carry on all the processes necessary for life. But it is left again in an alien land world when the last of these highest tides of the fortnight ebbs away; then its only defense is the firm closing of the plates of its shell to hold some of the moisture of the sea about its body. In its life brief and intense activity alternates with long periods of a quiescent state resembling hibernation. Like the plants of the Arctic, which must crowd the making and storing of food, the putting forth of flowers, and the forming of seeds into a few brief weeks of summer, this barnacle has drastically adjusted its way of life so that it may survive in a region of harsh conditions.

Some few sea animals have pushed on even above high water of the spring tides into the splash zone, where the only salty moisture comes from the spray of breaking waves. Among such pioneers are snails of the periwinkle tribe. One of the West Indian species can endure months of separation from the sea. Another, the European rock periwinkle, waits for the waves of the spring tides to cast its eggs into the sea, in almost all activities except the vital one of reproduction being independent of the water.

Below the low water of neaps are the areas exposed only as the rhythmic swing of the tides falls lower and lower, approaching the level of the springs. Of all the intertidal zone this region is linked most closely with the sea. Many of its inhabitants are offshore forms, able to live here only because of the briefness and infrequency of exposure to the air.

The relation between the tides and the zones of life is clear, but in many less obvious ways animals have adjusted their activities to the tidal rhythm. Some seem to be a mechanical matter of utilizing the movement of water. The larval oyster, for example, uses the flow of the tides to carry it into areas favorable for its attachment. Adult oysters live in bays or sounds or river estuaries rather than in water of full oceanic salinity, and so it is to the advantage of the race for the dispersal of the young stages to take place in a direction away from the open sea. When first hatched the larvae drift passively, the tidal currents carrying them now toward the sea, now toward the headwaters of estuaries or bays. In many estuaries the ebb tide runs longer than the flood, having the added push and volume of stream discharge behind it, and the resulting seaward drift over the whole two-week period of larval life would carry the young oysters many miles to sea. A sharp change of behavior sets in, however, as the larvae grow older. They now drop to the bottom while the tide ebbs, avoiding the seaward drift of water, but with the return of the flood they rise into the currents that are pressing upstream, and so are carried into regions of lower salinity that are favorable for their adult life.

Others adjust the rhythm of spawning to protect their young from the danger of being carried into unsuitable waters. One of the tube-building worms living in or near the tidal zone follows a pattern that avoids the strong movements of the spring tides. It releases its larvae into the sea every fortnight on the neap tides, when the water movements are relatively sluggish; the young worms, which have a very brief swimming stage, then have a good chance of remaining within the most favorable zone of the shore.

There are other tidal effects, mysterious and intangible. Sometimes spawning is synchronized with the tides in a way that suggests response to change of pressure or to the difference between still and flowing water. A primitive mollusk called the chiton spawns in Bermuda when the low tide occurs early in the morning, with the return flow of water setting in just after sunrise. As soon as the chitons are covered with water they shed their spawn. One of the Japanese nereid worms spawns only on the strongest tides of the year, near the new-and full-moon tides of October and November, presumably stirred in some obscure way by the amplitude of the water movements.

Many other animals, belonging to quite unrelated groups throughout the whole range of sea life, spawn according to a definitely fixed rhythm that may coincide with the full moon or the new moon or its quarters, but whether the effect is produced by the altered pressure of the tides or the changing light of the moon is by no means clear. For example, there is a sea urchin in Tortugas that spawns on the night of the full moon, and apparently only then. Whatever the stimulus may be, all the individuals of the species respond to it, assuring the simultaneous release of immense numbers of reproductive cells. On the coast of England one of the hydroids, an animal of plantlike appearance that produces tiny medusae or jellyfish, releases these medusae during the moon's third quarter. At Woods Hole on the Massachusetts coast a clamlike mollusk spawns heavily between the full and the new moon but avoids the first quarter. And a nereid worm at Naples gathers in its nuptial swarms during the quarters of the moon but never when the moon is new or full; a related worm at Woods Hole shows no such correlation although exposed to the same moon and to stronger tides.

In none of these examples can we be sure whether the animal is responding to the tides or, as the tides themselves do, to the influence of the moon. With plants, however, the situation is different, and here and there we find scientific confirmation of the ancient and world-wide belief in the effect of moonlight on vegetation. Various bits of evidence suggest that the rapid multiplication of diatoms and other members of the plant plankton is related to the phases of the moon. Certain algae in river plankton reach the peak of their abundance at the full moon. One of the brown seaweeds on the coast of North Carolina releases its reproductive cells only on the full moon, and similar behavior has been reported for other seaweeds in Japan and other parts of the world. These responses are generally explained as the effect of varying intensities of polarized light on protoplasm.

Other observations suggest some connection between plants and the reproduction and growth of animals. Rapidly maturing herring collect around the edge of concentrations of plant plankton, although the fully adult herring may avoid them. Spawning adults, eggs, and young of various other marine creatures are reported to occur more often in dense phytoplankton than in sparse patches. In significant experiments, a Japanese scientist discovered he could induce oysters to spawn with an extract obtained from sea lettuce. The same seaweed produces a substance that influences growth and multiplication of diatoms, and is itself stimulated by water taken from the vicinity of a heavy growth of rockweeds.

The whole subject of the presence in sea water of the so-called "ectocrines" (external secretions or products of metabolism) has so recently become one of the frontiers of science that actual information is fragmentary and tantalizing. It appears, however, that we may be on the verge of solving some of the riddles that have plagued men's minds for centuries. Though the subject lies in the misty borderlands of advancing knowledge, almost everything that in the past has been taken for granted, as well as problems considered insoluble, bear renewed thought in the light of the discovery of these substances.

In the sea there are mysterious comings and goings, both in space and time: the movements of migratory species, the strange phenomenon of succession by which, in one and the same area, one species appears in profusion, flourishes for a time, and then dies out, only to have its place taken by another and then another, like actors in a pageant passing before our eyes. And there are other mysteries. The phenomenon of "red tides" has been known from early days, recurring again and again down to the present time—a phenomenon in which the sea becomes discolored because of the extraordinary multiplication of some minute form, often a dinoflagellate, and in which there are disastrous side effects in the shape of mass mortalities among fish and some of the invertebrates. Then there is the problem of curious and seemingly erratic movements of fish, into or away from certain areas, often with sharp economic consequences. When the so-called "Atlantic water" floods the south coast of England, herring become abundant within the range of the Plymouth fisheries, certain characteristic plankton animals occur in profusion, and certain species of invertebrates flourish in the intertidal zone. When, however, this water mass is replaced by Channel water, the cast of characters undergoes many changes.

In the discovery of the biological role played by the sea water and all it contains, we may be about to reach an understanding of these old mysteries. For it is now clear that in the sea nothing lives to itself. The very water is altered, in its chemical nature and in its capacity for influencing life processes, by the fact that certain forms have lived within it and have passed on to it new substances capable of inducing far-reaching effects. So the present is linked with past and future, and each living thing with all that surrounds it.